BEYOND THE BASIC STUFF WITH PYTHON

BEYOND THE BASIC STUFF WITH PYTHON

Best Practices for Writing Clean Code

Al Sweigart

no starch press

San Francisco

Printed in the United States of America

First printing

24 23 22 21 20   1 2 3 4 5 6 7 8 9

ISBN-13: 978-1-59327-966-0 (print)
ISBN-13: 978-1-59327-967-7 (ebook)

Publisher: William Pollock
Executive Editor: Barbara Yien
Production Editor: Maureen Forys, Happenstance Type-O-Rama
Developmental Editor: Frances Saux
Cover Design: Octopod Studios
Interior Design: Octopod Studios
Cover Illustration: Josh Ellingson
Technical Reviewer: Kenneth Love
Copyeditor: Anne Marie Walker
Compositor: Happenstance Type-O-Rama
Proofreader: Rachel Monaghan
Indexer: Valerie Perry

For information on book distributors or translations, please contact No Starch Press, Inc. directly:
No Starch Press, Inc.
245 8th Street, San Francisco, CA 94103
phone: 1-415-863-9900; info@nostarch.com
www.nostarch.com

Library of Congress Cataloging-in-Publication Data

Library of Congress Cataloging-in-Publication Data

Names: Sweigart, Al, author.
Title: Beyond the basic stuff with python : best practices for writing clean code /
    Al Sweigart.
Description: San Francisco, CA : No Starch Press, Inc., [2021] | Includes
    index.
Identifiers: LCCN 2020034287 (print) | LCCN 2020034288 (ebook) | ISBN
    9781593279660 (paperback) | ISBN 9781593279677 (ebook)
Subjects: LCSH: Python (Computer program language) | Computer programming.
Classification: LCC QA76.73.P98 S943 2021 (print) | LCC QA76.73.P98
    (ebook) | DDC 005.13/3–dc23
LC record available at https://lccn.loc.gov/2020034287
LC ebook record available at https://lccn.loc.gov/2020034288

For my nephew Jack

About the Author

Al Sweigart is a software developer and tech book author living in Seattle. Python is his favorite programming language, and he is the developer of several open source modules for it. His other books are freely available under a Creative Commons license on his website at *https://www.inventwithpython.com/*. His cat Zophie weighs 11 pounds.

About the Technical Reviewer

Kenneth Love is a programmer, teacher, and conference organizer. He is a Django contributor and PSF Fellow, and currently works as a tech lead and software engineer for O'Reilly Media.

BRIEF CONTENTS

CONTENTS IN DETAIL

PART III: BEST PRACTICES, TOOLS, AND TECHNIQUES 43

3
CODE FORMATTING WITH BLACK 45

4
CHOOSING UNDERSTANDABLE NAMES 59

8

COMMON PYTHON GOTCHAS   **133**

9

ESOTERIC PYTHON ODDITIES   **153**

10

WRITING EFFECTIVE FUNCTIONS   **161**

11
COMMENTS, DOCSTRINGS, AND TYPE HINTS   181

12
ORGANIZING YOUR CODE PROJECTS WITH GIT   199

16
OBJECT-ORIENTED PROGRAMMING AND INHERITANCE   293

17
PYTHONIC OOP: PROPERTIES AND DUNDER METHODS   315

INDEX   339

ACKNOWLEDGMENTS

It's misleading to have just my name on the cover. This book wouldn't exist without the efforts of many people. I'd like to thank my publisher, Bill Pollock; and my editors, Frances Saux, Annie Choi, Meg Sneeringer, and Jan Cash. I'd like to also thank production editor Maureen Forys, copy editor Anne Marie Walker, and No Starch Press executive editor Barbara Yien. Thanks to Josh Ellingson for another great cover illustration. Thank you to my technical reviewer, Kenneth Love, and all the other great friends I've met in the Python community.

INTRODUCTION

Hello again, world! As a teenage programmer and wannabe hacker in the late 1990s, I would pore over the latest issues of *2600: The Hacker Quarterly*. One day, I finally summoned the courage to attend the magazine's monthly meetup in my city and was in awe of how knowledgeable everyone else seemed. (Later, I'd realize that many of them had more confidence than actual knowledge.) I spent the entire meeting nodding along to what others were saying, trying to keep up with their conversations. I left that meetup determined to spend every waking hour studying computing, programming, and network security so I could join the discussions at the next month's meetup.

At the next meetup, I continued to just nod and feel dumb compared to everyone else. So again I resolved to study and become "smart enough" to keep up. Month after month, I would increase my knowledge but always felt behind. I began to realize the enormity of the computing field and worried I would never know enough.

I knew more about programming than my high school friends but certainly not enough to get a job as a software developer. In the 1990s, Google, YouTube, and Wikipedia didn't exist. But even if those resources were available, I wouldn't have known how to use them; I wouldn't have been sure what to study next. Instead, I learned how to write *Hello, world!* programs in different programming languages but still felt I wasn't making real progress. I didn't know how to move beyond the basics.

There's so much more to software development than loops and functions. But once you've completed a beginner course or read an introductory programming book, your search for more guidance leads to yet another *Hello, world!* tutorial. Programmers often call this period the *desert of despair*: the time you spend wandering aimlessly through different learning materials, feeling like you're not improving. You become too advanced for beginner materials but too inexperienced to tackle more complex topics.

Those in this desert experience a strong sense of impostor syndrome. You don't feel like a "real" programmer or know how to craft code the way "real" programmers do. I wrote this book to address this audience. If you've learned the basics of Python, this book should help you become a more capable software developer and lose this sense of despair.

Who Should Read This Book and Why

This book targets those who have completed a basic Python tutorial and want to know more. The tutorial you learned from could have been my previous book, *Automate the Boring Stuff with Python* (No Starch Press, 2019), a book such as *Python Crash Course* (No Starch Press, 2019) by Eric Matthes, or an online course.

These tutorials might have hooked you on programming, but you still need more skills. If you feel like you're not yet at the professional programmer level but don't know how to get to that level, this is the book for you.

Or perhaps you were introduced to programming via another language besides Python and you want to jump right in to Python and its ecosystem of tools without retreading the same *Hello, world!* basics. If so, you don't need to read hundreds of pages that explain basic syntax; instead, skimming the "Learn Python in Y Minutes" article at *https://learnxinyminutes.com/docs/python/* or Eric Matthes's "Python Crash Course—Cheat Sheet" page at *https://ehmatthes.github.io/pcc/cheatsheets/README.html* will suffice before you tackle this book.

About This Book

This book covers more than just deeper-level Python syntax. It also discusses using the command line and the command line tools that professional developers use, such as code formatters, linters, and version control.

I explain what makes code readable and how you can write clean code. I've featured a few programming projects, so you can see these principles applied in actual software. Although this isn't a computer science textbook, I also explain Big O algorithm analysis and object-oriented design.

No single book can transform a person into a professional software developer, but I've written this book to further your knowledge toward that end. I introduce several topics that you might only otherwise discover, piecemeal, through hard-earned experience. After completing this book, your footing will be on a firmer foundation so you'll be better equipped to take on new challenges.

Although I recommend you read the chapters in this book in order, feel free to skip to whichever chapters capture your interest:

Part I: Getting Started

Chapter 1: Dealing with Errors and Asking for Help   Shows you how to effectively ask questions and find answers on your own. It also teaches you how to read error messages and the etiquette for asking for help online.

Chapter 2: Environment Setup and the Command Line   Explains how to navigate the command line along with setting up your development environment and the PATH environment variable.

Part II: Best Practices, Tools, and Techniques

Chapter 3: Code Formatting with Black   Describes the PEP 8 style guide and how to format your code to make it more readable. You'll learn how to automate this process using the Black code-formatting tool.

Chapter 4: Choosing Understandable Names   Describes how you should name your variables and functions to improve code readability.

Chapter 5: Finding Code Smells   Lists several potential red flags that could indicate the existence of bugs in your code.

Chapter 6: Writing Pythonic Code   Details several ways to write idiomatic Python code and what makes for *Pythonic* code.

Chapter 7: Programming Jargon   Explains technical terms used in the programming field and terms that are commonly confused with each other.

Chapter 8: Common Python Gotchas   Covers common sources of confusion and bugs in the Python language and how to correct them, as well as coding strategies to avoid.

Chapter 9: Esoteric Python Oddities   Covers several odd quirks of the Python language, such as string interning and the antigravity Easter egg, that you might not otherwise notice. You'll get an advanced understanding of how Python works by figuring out why some data types and operators result in such unexpected behavior.

Chapter 10: Writing Effective Functions   Details how to structure your functions for the most utility and readability. You'll learn about the * and ** argument syntax, the trade-offs between large and small functions, and functional programming techniques, such as lambda functions.

Chapter 11: Comments, Docstrings, and Type Hints   Covers the importance of the non-code parts of your program and how they affect maintainability. It includes how often you should write comments and docstrings, and how to make them informative. The chapter also discusses type hints and how to use static analyzers, such as Mypy, to detect bugs.

Chapter 12: Organizing Your Code Projects with Git   Describes using the Git version control tool to record the history of changes you make to your source code and recover previous versions of your work or track down when a bug first appeared. It also touches on how to structure your code projects' files using the Cookiecutter tool.

Chapter 13: Measuring Performance and Big O Algorithm Analysis Explains how to objectively measure your code's speed using the timeit and cProfile modules. In addition, it covers Big O algorithm analysis and how it lets you predict the way your code's performance slows down as the amount of data it has to process grows.

Chapter 14: Practice Projects   Has you apply the techniques you learned in this part by writing two command line games: the Tower of Hanoi, a puzzle game involving moving disks from one tower to the next, and the classic Four-in-a-Row board game for two players.

Part III: Object-Oriented Python

Chapter 15: Object-Oriented Programming and Classes   Defines the role of object-oriented programming (OOP) because it's often misunderstood. Many developers overuse OOP techniques in their code because they believe it's what everyone else does, but this leads to complicated source code. This chapter teaches you how to write classes, but more important, it teaches you why you should and shouldn't use them.

Chapter 16: Object-Oriented Programming and Inheritance   Explains class inheritance and its utility for code reuse.

Chapter 17: Pythonic OOP: Properties and Dunder Methods   Covers the Python-specific features in object-oriented design, such as properties, dunder methods, and operator overloading.

Your Programming Journey

The journey from novice to capable programmer can often feel like attempting to drink from a fire hose. With so many resources to choose from, you might worry that you're wasting time on suboptimal programming guides.

After you finish reading this book (or even while you're reading this book), I recommend following up by reading these additional introductory materials:

Python Crash Course (No Starch Press, 2019) by Eric Matthes is a book for beginners, but its project-based approach gives even experienced programmers a taste of Python's Pygame, matplotlib, and Django libraries.

Impractical Python Projects (No Starch Press, 2018) by Lee Vaughan provides a project-based approach to expand your Python skills. The programs you'll create by following the instructions in this book are fun and great programming practice.

Serious Python (No Starch Press, 2018) by Julien Danjou describes the steps you need to take to progress from a *garage project* hobbyist to a knowledgeable software developer who follows industry best practices and writes code that can scale.

But the technical aspects of Python are only one of its strengths. The programming language has attracted a diverse community responsible for creating a friendly, accessible body of documentation and support that no other programming ecosystem has matched. The annual PyCon conference, along with the many regional PyCons, hosts a wide variety of talks for all experience levels. The PyCon organizers make these talks available online for free at *https://pyvideo.org/*. The Tags page lets you easily find talks on topics that correspond to your interests.

To take a deeper dive into the advanced features of Python's syntax and standard library, I recommend reading the following titles:

Effective Python (Addison-Wesley Professional, 2019) by Brett Slatkin is an impressive collection of *Pythonic* best practices and language features.

Python Cookbook (O'Reilly Media, 2013) by David Beazley and Brian K. Jones offers an extensive list of code snippets to upgrade any Python novice's repertoire.

Fluent Python (O'Reilly Media, 2021) by Luciano Ramalho is a masterwork for exploring the intricacies of the Python language, and although its near-800-page size might be intimidating, it's well worth the effort.

Good luck on your programming journey. Let's get started!

PART 1

GETTING STARTED

1

DEALING WITH ERRORS AND ASKING FOR HELP

Please don't anthropomorphize computers; they find it very annoying. When a computer presents you with an error message, it's not because you've offended it. Computers are the most sophisticated tools most of us will ever interact with, but still, they're just tools.

Even so, it's easy to blame these tools. Because much of learning to program is self-directed, it's common to feel like a failure when you still need to consult the internet multiple times a day, even though you've been studying Python for months. But even professional software developers search the internet or consult documentation to answer their programming questions.

Unless you have the financial or social resources to hire a private tutor who can answer your programming questions, you're stuck with your computer, internet search engines, and your own fortitude. Fortunately, your questions have almost certainly been asked before. As a programmer, being able to find answers on your own is far more important than any algorithms or data structure knowledge. This chapter guides you through developing this crucial skill.

How to Understand Python Error Messages

When they're confronted with an error message's large wall of technobabble text, many programmers' first impulse is to completely ignore it. But inside this error message is the answer to what's wrong with your program. Finding this answer is a two-step process: examining the traceback and doing an internet search of the error message.

Examining Tracebacks

Python programs crash when the code raises an exception that an except statement doesn't handle. When this happens, Python displays the exception's message and a *traceback*. Also called a *stack trace,* the traceback shows the place in your program where the exception happened and the trail of function calls that led up to it.

To practice reading tracebacks, enter the following buggy program and save it as *abcTraceback.py*. The line numbers are for reference only and aren't part of the program.

```
1. def a():
2.      print('Start of a()')
❶ 3.      b()  # Call b().
4.
5. def b():
6.      print('Start of b()')
❷ 7.      c()  # Call c().
8.
9. def c():
10.      print('Start of c()')
❸ 11.      42 / 0  # This will cause a zero divide error.
12.
13. a()  # Call a().
```

In this program, the a() function calls b() ❶, which calls c() ❷. Inside c(), the 42 / 0 expression ❸ causes a zero divide error. When you run this program, the output should look like this:

```
Start of a()
Start of b()
Start of c()
Traceback (most recent call last):
  File "abcTraceback.py", line 13, in <module>
    a()  # Call a().
  File "abcTraceback.py", line 3, in a
    b()  # Call b().
  File "abcTraceback.py", line 7, in b
    c()  # Call c().
```

```
File "abcTraceback.py", line 11, in c
    42 / 0  # This will cause a zero divide error.
ZeroDivisionError: division by zero
```

Let's examine this traceback line by line, starting with this line:

```
Traceback (most recent call last):
```

This message lets you know that what follows is a traceback. The most recent call last text indicates that each of the function calls is listed in order, starting with the first function call and ending with the most recent. The next line shows the traceback's first function call:

```
File "abcTraceback.py", line 13, in <module>
    a()  # Call a().
```

These two lines are the *frame summary*, and they show the information inside a frame object. When a function is called, the local variable data as well as where in the code to return to after the function call ends are stored in a *frame object*. Frame objects hold local variables and other data associated with function calls. Frame objects are created when the function is called and destroyed when the function returns. The traceback shows a frame summary for each frame leading up to the crash. We can see that this function call is on line 13 of *abcTraceback.py*, and the <module> text tells us this line is in the global scope. Line 13 is displayed with two spaces of indentation next.

The four lines that follow are the next two frame summaries:

```
File "abcTraceback.py", line 3, in a
    b()  # Call b().
File "abcTraceback.py", line 7, in b
    c()  # Call c().
```

We can tell from the line 3, in a text that b() was called on line 3 inside the a() function, which led to c() being called on line 7 inside the b() function. Notice that the print() calls on lines 2, 6, and 10 aren't displayed in the traceback, even though they ran before the function calls occurred. Only the lines containing function calls that lead up to the exception are displayed in the traceback.

The last frame summary shows the line that caused the unhandled exception, followed by the name of the exception and the exception's message:

```
File "abcTraceback.py", line 11, in c
    42 / 0  # This will cause a zero divide error.
ZeroDivisionError: division by zero
```

Note that the line number given by the traceback is where Python finally detected an error. The true source of the bug could be somewhere before this line.

Error messages are notoriously short and inscrutable: the three words division by zero won't mean anything to you unless you know that dividing a number by zero is mathematically impossible and a common software bug. In this program, the bug isn't too hard to find. Looking at the line of code in the frame summary, it's clear where in the 42 / 0 code the zero divide error is happening.

But let's look at a more difficult case. Enter the following code into a text editor and save it as *zeroDivideTraceback.py*:

```
def spam(number1, number2):
    return number1 / (number2 - 42)

spam(101, 42)
```

When you run this program, the output should look like this:

```
Traceback (most recent call last):
  File "zeroDivideTraceback.py", line 4, in <module>
    spam(101, 42)
  File "zeroDivideTraceback.py", line 2, in spam
    return number1 / (number2 - 42)
ZeroDivisionError: division by zero
```

The error message is the same, but the zero divide in return number1 / (number2 - 42) isn't quite so obvious. You can deduce that there is a division happening from the / operator, and that the expression (number2 - 42) must evaluate to 0. This would lead you to conclude that the spam() function fails whenever the number2 parameter is set to 42.

Sometimes the traceback might indicate that an error is on the line after the true cause of the bug. For example, in the following program, the first line is missing the closing parenthesis:

```
print('Hello.'
print('How are you?')
```

But the error message for this program indicates the problem is on the second line:

```
  File "example.py", line 2
    print('How are you?')
        ^
SyntaxError: invalid syntax
```

The reason is that the Python interpreter didn't notice the syntax error until it read the second line. The traceback can indicate where things went wrong, but that isn't always the same as where the actual cause of a bug is. If the frame summary doesn't give you enough information to figure out the bug, or if the true cause of the bug is on a previous line not shown by the traceback, you'll have to step through the program with a debugger or check any logging messages to find the cause. This can take a significant amount of time. An internet search of the error message might give you critical clues about the solution much more quickly.

Searching for Error Messages

Often, error messages are so short they're not even full sentences. Because programmers encounter them regularly, they're intended as reminders rather than full explanations. If you're encountering an error message for the first time, copying and pasting it into an internet search frequently returns a detailed explanation of what the error means and what its likely causes are. Figure 1-1 shows the results of a search for **python "ZeroDivisionError: division by zero"**. Including quotation marks around the error message helps find the exact phrase, and adding the word *python* can narrow down your search as well.

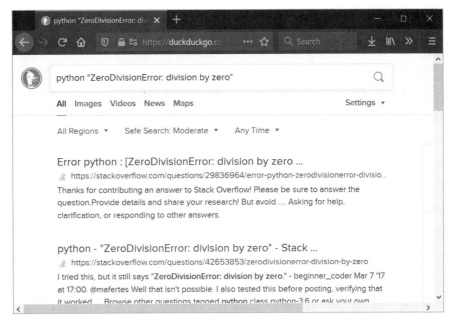

Figure 1-1: Copying and pasting an error message into an internet search tool can quickly provide explanations and solutions.

Searching for error messages isn't cheating. Nobody can be expected to memorize every possible error message for a programming language. Professional software developers search the internet for programming answers on a daily basis.

You might want to exclude any part of the error message that is particular to your code. For example, consider the following error messages:

```
>>> print(employeRecord)
Traceback (most recent call last):
  File "<stdin>", line 1, in <module>
```

❶ NameError: name 'employeeRecord' is not defined
```
>>> 42 - 'hello'
Traceback (most recent call last):
  File "<stdin>", line 1, in <module>
```
❷ TypeError: unsupported operand type(s) for -: 'int' and 'str'

This example has a typo in the variable employeeRecord, causing an error ❶. Because the identifier employeeRecord in NameError: name 'employeeRecord' is not defined is specific to your code, you might want to instead search for **python "NameError: name" "is not defined"**. In the last line, the 'int' and 'str' part of the error message ❷ seems to refer to the 42 and 'hello' values, so truncating the search to **python "TypeError: unsupported operand type(s) for"** would avoid including parts particular to your code. If these searches don't yield useful results, try including the full error message.

Preventing Errors with Linters

The best way to fix mistakes is to not make them in the first place. Lint software, or *linters*, are applications that analyze your source code to warn you of any potential errors. The name references the small fibers and debris collected by a clothes dryer's lint trap. Although a linter won't catch all errors, *static analysis* (examining source code without running it) can identify common errors caused by typos. (Chapter 11 explores how to use type hints for static analysis.) Many text editors and integrated development environments (IDEs) incorporate a linter that runs in the background and can point out problems in real time, such as in Figure 1-2.

Figure 1-2: A linter points out an undefined variable in Mu (top), PyCharm (middle), and Sublime Text (bottom).

The near-instant notifications that a linter provides greatly improves your programming productivity. Without one, you'd have to run your program, watch it crash, read the traceback, and then find the line in your source code to fix a typo. And if you've made multiple typos, this run-fix cycle would only find them one at a time. Linting can point out multiple errors at once, and it does so directly in the editor, so you can see the line on which they occur.

Your editor or IDE might not come with a lint feature, but if it supports plug-ins, almost certainly a linter will be available. Often, these plug-ins use a linting module called Pyflakes or some other module to do their analysis. You can install Pyflakes from *https://pypi.org/project/pyflakes/* or by running `pip install --user pyflakes`. It's well worth the effort.

NOTE *On Windows, you can run the `python` and `pip` commands. But on macOS and Linux, these command names are for Python version 2 only, so instead you'll need to run `python3` and `pip3`. Keep this in mind whenever you see `python` or `pip` in this book.*

IDLE, the IDE that comes with Python, doesn't have a linter or the capability of having one installed.

How to Ask for Programming Help

When internet searches and linters fail to solve your problem, you can ask for programming help on the internet. But there is an etiquette to efficiently asking for advice. If experienced software developers are willing to answer your questions at no charge, it's best to make efficient use of their time.

Asking strangers for programming help should always be a last resort. Hours or days could pass before someone replies to your posted question, if you get a reply at all. It's much faster to search the web for other people who have already asked your question and read their answers. Online documentation and search engines were made to relieve the question-answering work that would otherwise have to be done by humans.

But when you've exhausted your options and must ask a human audience your programming question, avoid the following common mistakes:

- Asking if it's okay to ask a question instead of just asking it
- Implying your question instead of asking it directly
- Asking your question on the wrong forum or website
- Writing an unspecific post headline or email subject, such as "I have a problem" or "Please help"
- Saying "my program doesn't work" but not explaining how you want it to work
- Not including the full error message
- Not sharing your code

- Sharing poorly formatted code
- Not explaining what you've already tried
- Not giving operating system or version information
- Asking someone to write a program for you

This list of "don'ts" isn't just for decorum; these habits prevent your helpers from helping you. Your helper's first step will be to run your code and try to reproduce your problem. To do so, they'll need a lot of information about your code, computer, and intentions. It's far more common to provide too little information than too much. The next several sections explore what you can do to prevent these common mistakes. I'll assume that you're posting your question to an online forum, but these guidelines also apply to cases when you're emailing questions to a single person or mailing list.

Limit Back and Forth by Providing Your Information Upfront

If you approached someone in person, asking "Can I ask you a question?" would be a short, pleasant means to see if your helper was available. But on online forums, your helper can hold off on a reply until they have the time to do so. Because there could be hours between replies, it's best to supply all the information your helper might need in your initial post instead of asking for permission to ask your question. If they don't reply, you can copy and paste this information to a different forum.

State Your Question in the Form of an Actual Question

It's easy to assume that your helpers know what you're talking about when you explain your problem. But programming is an expansive field, and they might not have experience with the particular area in which you're having trouble. So it's important to state your question in the form of an actual question. Although sentences that begin with "I want to . . ." or "The code isn't working" can imply what your question is, be sure to include explicit questions: literally, sentences that end with a question mark. Otherwise, it's probably unclear what you're asking.

Ask Your Question on the Appropriate Website

Asking a Python question on a JavaScript forum or an algorithms question on a network security mailing list will likely be unproductive. Often, mailing lists and online forums have Frequently Asked Questions (FAQ) documents or description pages that explain which topics are appropriate to discuss. For example, the *python-dev* mailing list is about the Python language's design features, so it isn't a general Python help mailing list. The web page at *https://www.python.org/about/help/* can direct you to an appropriate place to ask whatever sort of Python question you have.

Summarize Your Question in the Headline

The benefit of posting your question to an online forum is that future programmers who have the same question can find it and its answers using an internet search. Be sure to use a headline that summarizes the question to make it easy for search engines to organize. A generic headline like "Help please" or "Why isn't this working?" is too vague. If you're asking your question in an email, a meaningful subject line tells your helper what your question is as they scan their inbox.

Explain What You Want the Code to Do

The question "Why doesn't my program work?" omits the critical detail of what you want your program to do. This isn't always obvious to your helper, because they don't know what your intention is. Even if your question is just "Why am I getting this error?" it helps to also say what your program's end goal is. In some cases, your helper can tell you if you need an entirely different approach, and you can abandon your problem rather than wasting time trying to solve it.

Include the Full Error Message

Be sure to copy and paste the entire error message, including the traceback. Merely describing your error, such as "I'm getting an out of range error," doesn't provide enough detail for your helper to figure out what is wrong. Also, specify whether you always encounter this error or if it's an intermittent problem. If you've identified the specific circumstances in which the error happens, include those details as well.

Share Your Complete Code

Along with the full error message and traceback, provide the source code for your entire program. That way, your helper can run your program on their machine under a debugger to examine what is happening. Always produce a *minimum, complete, and reproducible* (*MCR*) example that reliably reproduces the error you're getting. The MCR term comes from Stack Overflow and is discussed in detail at *https://stackoverflow.com/help/mcve/*. *Minimal* means your code example is as short as possible while still reproducing the problem you're encountering. *Complete* means that your code example contains everything it needs to reproduce the problem. *Reproducible* means that your code example reliably reproduces the problem you're describing.

But if your program is contained in one file, sending it to your helper is a simple matter. Just ensure that it's properly formatted, as discussed in the next section.

Make Your Code Readable with Proper Formatting

The point of sharing your code is so your helper can run your program and reproduce the error you're getting. Not only do they need the code, but they also need it properly formatted. Make sure that they can easily copy your source and run it as is. If you're copying and pasting your source code in an email, be aware that many email clients might remove the indentation, resulting in code that looks like this:

```
def knuts(self, value):
if not isinstance(value, int) or value < 0:
raise WizCoinException('knuts attr must be a positive int')
self._knuts = value
```

Not only would it take a long time for your helper to reinsert the indentation for every line in your program, but it's ambiguous as to how much indentation each line had to begin with. To ensure your code is properly formatted, copy and paste your code to a *pastebin* website, such as *https://pastebin.com/* or *https://gist.github.com/*, which stores your code at a short, public URL, such as *https://pastebin.com/XeU3yusC*. Sharing this URL is easier than using a file attachment.

If you're posting code to a website, such as *https://stackoverflow.com/* or *https://reddit.com/r/learnpython/*, make sure you use the formatting tools its text boxes provide. Often, indenting a line with four spaces will ensure that line uses a monospace "code font," which is easier to read. You can also enclose text with a backtick (`) character to put it in the monospace code font. These sites frequently have a link to formatting information. Not using these tips might mangle your source code, making it all appear on one line, like the following:

```
def knuts(self, value):if not isinstance(value, int) or value < 0:raise
WizCoinException('knuts attr must be a positive int') self._knuts = value
```

In addition, don't share your code by taking a screenshot or a photo of your screen and sending the image. It's impossible to copy and paste the code from the image, and it's usually unreadable as well.

Tell Your Helper What You've Already Tried

When posting your question, tell your helper what you've already tried and the results of those tries. This information saves your helper the effort of retrying these false leads and shows that you've put effort into solving your own problem.

Additionally, this information ensures that you're asking for help, not just asking for someone to write your software for you. Unfortunately, it's common for computer science students to ask online strangers to do their homework or for entrepreneurs to ask for someone to create a "quick app" for them for free. Programming help forums aren't made for this purpose.

Describe Your Setup

Your computer's particular setup might affect how your program runs and what errors it produces. To ensure that your helpers can reproduce your problem on their computer, give them the following information about your computer:

- The operating system and version, such as "Windows 10 Professional Edition" or "macOS Catalina"
- The Python version running the program, such as "Python 3.7" or "Python 3.6.6"
- Any third-party modules your program uses and their versions, such as "Django 2.1.1"

You can find the versions of your installed third-party modules by running pip list. It's also a convention to include the module's version in the __version__ attribute, as in the following interactive shell example:

```
>>> import django
>>> django.__version__
'2.1.1'
```

Most likely, this information won't be necessary. But to reduce the back and forth, offer this information in your initial post anyway.

Examples of Asking a Question

Here is a properly asked question that follows the dos and don'ts of the previous section:

> **Selenium webdriver: How do I find ALL of an element's attributes?**
>
> In the Python Selenium module, once I have a WebElement object I can get the value of any of its attributes with get_attribute():
>
> foo = elem.get_attribute('href')
>
> If the attribute named 'href' doesn't exist, None is returned.
>
> My question is, how can I get a list of all the attributes that an element has? There doesn't seem to be a get_attributes() or get_attribute_names() method.
>
> I'm using version 2.44.0 of the Selenium module for Python.

This question comes from *https://stackoverflow.com/q/27307131/1893164/*. The headline summarizes the question in a single sentence. The problem is stated in the form of a question and ends with a question mark. In the future, if a person reads this headline in an internet search result, they'll immediately know whether or not it's relevant to their own question.

The question formats the code using the monospace code font and breaks up text across multiple paragraphs. It's clear what the question in this post is: it's even prefaced with "My question is." It suggests that get_attributes() or get_attribute_names() could have been, but aren't, answers, which shows that the asker has tried to find a solution while hinting at what they believe the true answer to this question would look like. The asker also includes the Selenium module's version information just in case it's relevant. It's better to include too much information than not enough.

Summary

Independently answering your own programming questions is the most important skill a programmer must learn. The internet, which was built by programmers, has a wealth of resources that provide the answers you need.

But first, you must parse the often cryptic error messages that Python raises. It's fine if you can't understand the text of an error message. You can still submit this text to a search engine to find the error message's plain English explanation and the likely cause. The error's traceback will indicate where in your program the error occurred.

A real-time linter can point out typos and potential bugs as you write code. Linters are so useful that modern software development effectively requires them. If your text editor or IDE doesn't have a linter or the ability to add a linter plug-in, consider switching to one that does.

If you can't find the solution to your problem by searching the internet, try posting your question to an online forum or email someone. To make this process efficient, this chapter provided guidelines for asking a good programming question. This includes asking a specific, well-stated question, providing full source code and error message details, explaining what you've already tried, and telling your helper which operating system and Python version you're using. Not only will the posted answers solve your problem, but they can help future programmers who have the same question and find your post.

Don't feel discouraged if you seem to be constantly looking up answers and asking for help. Programming is an extensive field, and no one can hold all of its details in their head at once. Even experienced software developers check online for documentation and solutions daily. Instead, focus on becoming skillful at finding solutions, and you'll be on your way to becoming a proficient Pythonista.

2

ENVIRONMENT SETUP
AND THE COMMAND LINE

Environment setup is the process of organizing your computer so you can write code. This involves installing any necessary tools, configuring them, and handling any hiccups during the setup. There is no single setup process because everyone has a different computer with a different operating system, version of the operating system, and version of the Python interpreter. Even so, this chapter describes some basic concepts to help you administer your own computer using the command line, environment variables, and filesystem.

Learning these concepts and tools might seem like a headache. You want to write code, not poke around configuration settings or understand inscrutable console commands. But these skills will save you time in the long run. Ignoring error messages or randomly changing configuration settings to get your system working well enough might hide problems, but it won't fix them. By taking the time to understand these issues now, you can prevent them from reoccurring.

The Filesystem

The *filesystem* is how your operating system organizes data to be stored and retrieved. A file has two key properties: a *filename* (usually written as one word) and a *path*. The path specifies the location of a file on the computer. For example, a file on my Windows 10 laptop has the filename *project.docx* in the path *C:\Users\Al\Documents*. The part of the filename after the last period is the file's *extension* and tells you a file's type. The filename *project.docx* is a Word document, and *Users*, *Al*, and *Documents* all refer to *folders* (also called *directories*). Folders can contain files and other folders. For example, *project.docx* is in the *Documents* folder, which is in the *Al* folder, which is in the *Users* folder. Figure 2-1 shows this folder organization.

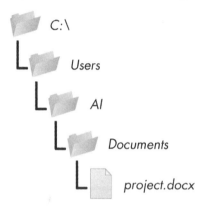

Figure 2-1: A file in a hierarchy of folders

The *C:* part of the path is the *root folder*, which contains all other folders. On Windows, the root folder is named *C:* and is also called the *C:* drive. On macOS and Linux, the root folder is */.* In this book, I'll use the Windows-style root folder, *C:\.* If you're entering the interactive shell examples on macOS or Linux, enter / instead.

Additional volumes, such as a DVD drive or USB flash drive, will appear differently on different operating systems. On Windows, they appear as new, lettered root drives, such as *D:* or *E:\.* On macOS, they appear as new folders within the */Volumes* folder. On Linux, they appear as new folders within the */mnt* ("mount") folder. Note that folder names and filenames are not case sensitive on Windows and macOS, but they're case sensitive on Linux.

Paths in Python

On Windows, the backslash (\) separates folders and filenames, but on macOS and Linux, the forward slash (/) separates them. Instead of writing code both ways to make your Python scripts cross-platform compatible, you can use the pathlib module and / operator instead.

The typical way to import pathlib is with the statement from pathlib import Path. Because the Path class is the most frequently used class in pathlib, this form lets you type Path instead of pathlib.Path. You can pass a string of a folder or filename to Path() to create a Path object of that folder or filename. As long as the leftmost object in an expression is a Path object, you can use the / operator to join together Path objects or strings. Enter the following into the interactive shell:

```
>>> from pathlib import Path
>>> Path('spam') / 'bacon' / 'eggs'
WindowsPath('spam/bacon/eggs')
>>> Path('spam') / Path('bacon/eggs')
WindowsPath('spam/bacon/eggs')
>>> Path('spam') / Path('bacon', 'eggs')
WindowsPath('spam/bacon/eggs')
```

Note that because I ran this code on a Windows machine, Path() returns WindowsPath objects. On macOS and Linux, a PosixPath object is returned. (POSIX is a set of standards for Unix-like operating systems and is beyond the scope of this book.) For our purposes, there's no difference between these two types.

You can pass a Path object to any function in the Python standard library that expects a filename. For example, the function call open(Path('C:\\') / 'Users' / 'Al' / 'Desktop' / 'spam.py') is equivalent to open(r'C:\Users\Al\Desktop\spam.py').

The Home Directory

All users have a folder called the *home folder* or *home directory* for their own files on the computer. You can get a Path object of the home folder by calling Path.home():

```
>>> Path.home()
WindowsPath('C:/Users/Al')
```

The home directories are located in a set place depending on your operating system:

- On Windows, home directories are in *C:\Users*.
- On Mac, home directories are in */Users*.
- On Linux, home directories are often in */home*.

Your scripts will almost certainly have permissions to read from and write to the files in your home directory, so it's an ideal place to store the files that your Python programs will work with.

The Current Working Directory

Every program that runs on your computer has a *current working directory* (*cwd*). Any filenames or paths that don't begin with the root folder you can assume are in the cwd. Although "folder" is the more modern name for a directory, note that cwd (or just working directory) is the standard term, not "current working folder."

You can get the cwd as a Path object using the `Path.cwd()` function and change it using `os.chdir()`. Enter the following into the interactive shell:

```
>>> from pathlib import Path
>>> import os
❶ >>> Path.cwd()
WindowsPath('C:/Users/Al/AppData/Local/Programs/Python/Python38')
❷ >>> os.chdir('C:\\Windows\\System32')
>>> Path.cwd()
WindowsPath('C:/Windows/System32')
```

Here, the cwd was set to *C:\Users\Al\AppData\Local\Programs\Python\Python38* ❶, so the filename *project.docx* would refer to *C:\Users\Al\AppData\Local\Programs\Python\Python38\project.docx*. When we change the cwd to *C:\Windows\System32* ❷, the filename *project.docx* would refer to *C:\Windows\System32\project.docx*.

Python displays an error if you try to change to a directory that doesn't exist:

```
>>> os.chdir('C:/ThisFolderDoesNotExist')
Traceback (most recent call last):
  File "<stdin>", line 1, in <module>
FileNotFoundError: [WinError 2] The system cannot find the file specified:
'C:/ThisFolderDoesNotExist'
```

The `os.getcwd()` function in the os module is a former way of getting the cwd as a string.

Absolute vs. Relative Paths

There are two ways to specify a file path:

- An absolute path, which always begins with the root folder
- A relative path, which is relative to the program's cwd

There are also the *dot* (.) and *dot-dot* (..) folders. These are not real folders but special names that you can use in a path. A single period (.) for a folder name is shorthand for "this directory." Two periods (..) means "the parent folder."

Figure 2-2 shows an example of some folders and files. When the cwd is set to *C:\bacon*, the relative paths for the other folders and files are set as they are in the figure.

The .\ at the start of a relative path is optional. For example, *.\spam.txt* and *spam.txt* refer to the same file.

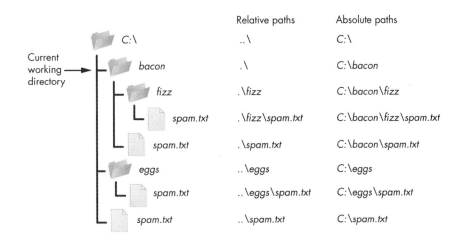

	Relative paths	Absolute paths
C:\	..\	C:\
bacon	.\	C:\bacon
fizz	.\fizz	C:\bacon\fizz
spam.txt	.\fizz\spam.txt	C:\bacon\fizz\spam.txt
spam.txt	.\spam.txt	C:\bacon\spam.txt
eggs	..\eggs	C:\eggs
spam.txt	..\eggs\spam.txt	C:\eggs\spam.txt
spam.txt	..\spam.txt	C:\spam.txt

Current working directory → bacon

Figure 2-2: The relative paths for folders and files in the working directory C:\bacon

Programs and Processes

A *program* is any software application that you can run, such as a web browser, spreadsheet application, or word processor. A *process* is a running instance of a program. For example, Figure 2-3 shows five running processes of the same calculator program.

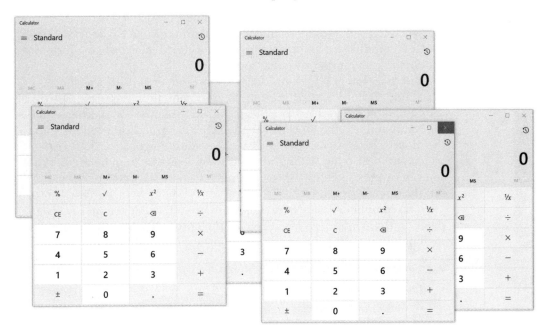

Figure 2-3: One calculator program running multiple times as multiple, separate processes

Processes remain separate from each other, even when running the same program. For example, if you ran several instances of a Python program at the same time, each process might have separate variable values. Every process, even processes running the same program, has its own cwd and environment variable settings. Generally speaking, a command line will run only one process at a time (although you can have multiple command lines open simultaneously).

Each operating system has a way of viewing a list of running processes. On Windows, you can press CTRL-SHIFT-ESC to bring up the Task Manager application. On macOS, you can run Applications ▸ Utilities ▸ Activity Monitor. On Ubuntu Linux, you can press CTRL-ALT-DEL to open an application also called the Task Manager. These task managers can force a running process to terminate if it's unresponsive.

The Command Line

The *command line* is a text-based program that lets you enter commands to interact with the operating system and run programs. You might also hear it called the command line interface (CLI, which rhymes with "fly"), command prompt, terminal, shell, or console. It provides an alternative to a *graphical user interface* (*GUI*, pronounced "gooey"), which allows the user to interact with the computer through more than just a text-based interface. A GUI presents visual information to a user to guide them through tasks more easily than the command line does. Most computer users treat the command line as an advanced feature and never touch it. Part of the intimidation factor is due to the complete lack of hints of how to use it; although a GUI might display a button showing you where to click, a blank terminal window doesn't remind you what to type.

But there are good reasons for becoming adept at using the command line. For one, setting up your environment often requires you to use the command line rather than the graphical windows. For another, entering commands can be much faster than clicking graphical windows with the mouse. Text-based commands are also less ambiguous than dragging an icon to some other icon. This lends them to automation better, because you can combine multiple specific commands into scripts to perform sophisticated operations.

The command line program exists in an executable file on your computer. In this context, we often call it a shell or shell program. Running the shell program makes the terminal window appear:

- On Windows, the shell program is at *C:\Windows\System32\cmd.exe*.
- On macOS, the shell program is at */bin/bash*.
- On Ubuntu Linux, the shell program is at */bin/bash*.

Over the years, programmers have created many shell programs for the Unix operating system, such as the Bourne Shell (in an executable file named *sh*) and later the Bourne-Again Shell (in an executable file named *Bash*). Linux uses Bash by default, whereas macOS uses the similar

Zsh or Z shell in Catalina and later versions. Due to its different development history, Windows uses a shell named Command Prompt. All these programs do the same thing: they present a terminal window with a text-based CLI into which the user enters commands and runs programs.

In this section, you'll learn some of the command line's general concepts and common commands. You could master a large number of cryptic commands to become a real sorcerer, but you only need to know about a dozen or so to solve most problems. The exact command names might vary slightly on different operating systems, but the underlying concepts are the same.

Opening a Terminal Window

To open a terminal window, do the following:

- On Windows, click the Start button, type `Command Prompt`, and then press ENTER.
- On macOS, click the `Spotlight` icon in the upper-right corner, type `Terminal`, and then press ENTER.
- On Ubuntu Linux, press the WIN key to bring up Dash, type `Terminal`, and press ENTER. Alternatively, use the keyboard shortcut CTRL-ALT-T.

Like the interactive shell, which displays a >>> prompt, the terminal displays a *shell prompt* at which you can enter commands. On Windows, the prompt will be the full path to the current folder you are in:

```
C:\Users\Al>your commands go here
```

On macOS, the prompt shows your computer's name, a colon, and the cwd with your home folder represented as a tilde (~). After this is your username followed by a dollar sign ($):

```
Als-MacBook-Pro:~ al$ your commands go here
```

On Ubuntu Linux, the prompt is similar to the macOS prompt except it begins with the username and an at (@) symbol:

```
al@al-VirtualBox:~$ your commands go here
```

Many books and tutorials represent the command line prompt as just $ to simplify their examples. It's possible to customize these prompts, but doing so is beyond the scope of this book.

Running Programs from the Command Line

To run a program or command, enter its name into the command line. Let's run the default calculator program that comes with the operating system. Enter the following into the command line:

- On Windows, enter `calc.exe`.

- On macOS, enter **open -a Calculator**. (Technically, this runs the open program, which then runs the Calculator program.)
- On Linux, enter **gnome-calculator**.

Program names and commands are case sensitive on Linux but case insensitive on Windows and macOS. This means that even though you must type gnome-calculator on Linux, you could type `Calc.exe` on Windows and `OPEN -a Calculator` on macOS.

Entering these calculator program names into the command line is equivalent to running the Calculator program from the Start menu, Finder, or Dash. These calculator program names work as commands because the *calc.exe*, *open*, and *gnome-calculator* programs exist in folders that are included in the PATH environment variables. "Environment Variables and PATH" on page 35 explains this further. But suffice it to say that when you enter a program name on the command line, the shell checks whether a program with that name exists in one of the folders listed in PATH. On Windows, the shell looks for the program in the cwd (which you can see in the prompt) before checking the folders in PATH. To tell the command line on macOS and Linux to first check the cwd, you must enter ./ before the filename.

If the program isn't in a folder listed in PATH, you have two options:

- Use the cd command to change the cwd to the folder that contains the program, and then enter the program name. For example, you could enter the following two commands:

```
cd C:\Windows\System32
calc.exe
```

- Enter the full file path for the executable program file. For example, instead of entering calc.exe, you could enter C:\Windows\System32\calc.exe.

On Windows, if a program ends with the file extension *.exe* or *.bat*, including the extension is optional: entering calc does the same thing as entering calc.exe. Executable programs in macOS and Linux often don't have file extensions marking them as executable; rather, they have the executable permission set. "Running Python Programs Without the Command Line" on page 39 has more information.

Using Command Line Arguments

Command line arguments are bits of text you enter after the command name. Like the arguments passed to a Python function call, they provide the command with specific options or additional directions. For example, when you run the command cd C:\Users, the C:\Users part is an argument to the cd command that tells cd to which folder to change the cwd. Or, when you run a Python script from a terminal window with the python yourScript.py command, the yourScript.py part is an argument telling the python program what file to look in for the instructions it should carry out.

Command line options (also called flags, switches, or simply options) are a single-letter or short-word command line arguments. On Windows, command line options often begin with a forward slash (/); on macOS and Linux, they begin with a single dash (-) or double dash (--). You already used the -a option when running the macOS command open -a Calculator. Command line options are often case sensitive on macOS and Linux but are case insensitive on Windows, and we separate multiple command line options with spaces.

Folders and filenames are common command line arguments. If the folder or filename has a space as part of its name, enclose the name in double quotes to avoid confusing the command line. For example, if you want to change directories to a folder called *Vacation Photos*, entering cd Vacation Photos would make the command line think you were passing two arguments, Vacation and Photos. Instead, you enter cd "Vacation Photos":

```
C:\Users\Al>cd "Vacation Photos"

C:\Users\Al\Vacation Photos>
```

Another common argument for many commands is --help on macOS and Linux and /? on Windows. These bring up information associated with the command. For example, if you run cd /? on Windows, the shell tells you what the cd command does and lists other command line arguments for it:

```
C:\Users\Al>cd /?
Displays the name of or changes the current directory.

CHDIR [/D] [drive:][path]
CHDIR [..]
CD [/D] [drive:][path]
CD [..]

  ..   Specifies that you want to change to the parent directory.

Type CD drive: to display the current directory in the specified drive.
Type CD without parameters to display the current drive and directory.

Use the /D switch to change current drive in addition to changing current
directory for a drive.
--snip--
```

This help information tells us that the Windows cd command also goes by the name chdir. (Most people won't type chdir when the shorter cd command does the same thing.) The square brackets contain optional arguments. For example, CD [/D] [drive:][path] tells you that you could specify a drive or path using the /D option.

Unfortunately, although the /? and --help information for commands provides reminders for experienced users, the explanations can often be cryptic. They're not good resources for beginners. You're better off using a book or web tutorial instead, such as *The Linux Command Line,* 2nd Edition (2019) by William Shotts, *Linux Basics for Hackers* (2018) by OccupyTheWeb, or *PowerShell for Sysadmins* (2020) by Adam Bertram, all from No Starch Press.

Running Python Code from the Command Line with -c

If you need to run a small amount of throwaway Python code that you run once and then discard, pass the -c switch to python.exe on Windows or python3 on macOS and Linux. The code to run should come after the -c switch, enclosed in double quotes. For example, enter the following into the terminal window:

```
C:\Users\Al>python -c "print('Hello, world')"
Hello, world
```

The -c switch is handy when you want to see the results of a single Python instruction and don't want to waste time entering the interactive shell. For example, you could quickly display the output of the help() function and then return to the command line:

```
C:\Users\Al>python -c "help(len)"
Help on built-in function len in module builtins:

len(obj, /)
    Return the number of items in a container.

C:\Users\Al>
```

Running Python Programs from the Command Line

Python programs are text files that have the *.py* file extension. They're not executable files; rather, the Python interpreter reads these files and carries out the Python instructions in them. On Windows, the interpreter's executable file is *python.exe*. On macOS and Linux, it's *python3* (the original *python* file contains the Python version 2 interpreter). Running the commands python yourScript.py or python3 yourScript.py will run the Python instructions saved in a file named *yourScript.py*.

Running the py.exe Program

On Windows, Python installs a *py.exe* program in the *C:\Windows* folder. This program is identical to *python.exe* but accepts an additional command line argument that lets you run any Python version installed on your computer. You can run the py command from any folder, because the *C:\Windows* folder is included in the PATH environment variable. If you have multiple Python versions installed, running py automatically runs the latest version installed on your computer. You can also pass a -3 or -2 command line argument to run the latest Python version 3 or version 2 installed, respectively. Or you could enter a more specific version number, such as -3.6 or -2.7, to run that particular Python installation. After the version switch, you can pass all the same command line arguments to *py.exe* as you do to *python.exe*. Run the following from the Windows command line:

```
C:\Users\Al>py -3.6 -c "import sys;print(sys.version)"
3.6.6 (v3.6.6:4cf1f54eb7, Jun 27 2018, 03:37:03) [MSC v.1900 64 bit (AMD64)]

C:\Users\Al>py -2.7
Python 2.7.14 (v2.7.14:84471935ed, Sep 16 2017, 20:25:58) [MSC v.1500 64 bit
(AMD64)] on win32
Type "help", "copyright", "credits" or "license" for more information.
>>>
```

The *py.exe* program is helpful when you have multiple Python versions installed on your Windows machine and need to run a specific version.

Running Commands from a Python Program

Python's subprocess.run() function, found in the subprocess module, can run shell commands within your Python program and then present the command output as a string. For example, the following code runs the ls -al command:

```
>>> import subprocess, locale
❶>>> procObj = subprocess.run(['ls', '-al'], stdout=subprocess.PIPE)
❷>>> outputStr = procObj.stdout.decode(locale.getdefaultlocale()[1])
>>> print(outputStr)
total 8
drwxr-xr-x  2 al al 4096 Aug  6 21:37 .
drwxr-xr-x 17 al al 4096 Aug  6 21:37 ..
-rw-r--r--  1 al al    0 Aug  5 15:59 spam.py
```

We pass the ['ls', '-al'] list to subprocess.run() ❶. This list contains the command name ls, followed by its arguments, as individual strings. Note that passing ['ls -al'] wouldn't work. We store the command's output as a string in outputStr ❷. Online documentation for subprocess.run() and locale.getdefaultlocale() will give you a better idea of how these functions work, but they make the code work on any operating system running Python.

Minimizing Typing with Tab Completion

Because advanced users enter commands into computers for hours a day, modern command lines offer features to minimize the amount of typing necessary. The *tab completion* feature (also called command line completion or autocomplete) lets a user type the first few characters of a folder or filename and then press the TAB key to have the shell fill in the rest of the name.

For example, when you type cd c:\u and press TAB on Windows, the current command checks which folders or files in *C:* begin with *u* and tab completes to c:\Users. It corrects the lowercase *u* to *U* as well. (On macOS and Linux, tab completion doesn't correct the casing.) If multiple folders or filenames begin with *U* in the *C:*folder, you can continue to press TAB to cycle through all of them. To narrow down the number of matches, you could also type cd c:\us, which filters the possibilities to folders and filenames that begin with *us*.

Pressing the TAB key multiple times works on macOS and Linux as well. In the following example, the user typed cd D, followed by TAB twice:

```
al@al-VirtualBox:~$ cd D
Desktop/   Documents/ Downloads/
al@al-VirtualBox:~$ cd D
```

Pressing TAB twice after typing the D causes the shell to display all the possible matches. The shell gives you a new prompt with the command as you've typed it so far. At this point, you could type, say, e and then press TAB to have the shell complete the cd Desktop/ command.

Tab completion is so useful that many GUI IDEs and text editors include this feature as well. Unlike command lines, these GUI programs usually display a small menu under your words as you type them, letting you select one to autocomplete the rest of the command.

Viewing the Command History

In their *command history*, modern shells also remember the commands you've entered. Pressing the up arrow key in the terminal fills the command line with the last command you entered. You can continue to press the up arrow key to find earlier commands, or press the down arrow key to return to more recent commands. If you want to cancel the command currently in the prompt and start from a fresh prompt, press CTRL-C.

On Windows, you can view the command history by running doskey /history. (The oddly named *doskey* program goes back to Microsoft's pre-Windows operating system, MS-DOS.) On macOS and Linux, you can view the command history by running the history command.

Working with Common Commands

This section contains a short list of the common commands you'll use in the command line. There are far more commands and arguments than listed here, but you can treat these as the bare minimum you'll need to navigate the command line.

Command line arguments for the commands in this section appear between square brackets. For example, cd [destination folder] means you should enter cd, followed by the name of a new folder.

Match Folder and Filenames with Wildcard Characters

Many commands accept folder and filenames as command line arguments. Often, these commands also accept names with the wildcard characters * and ?, allowing you to specify multiple matching files. The * character matches any number of characters, whereas the ? character matches any single character. We call expressions that use the * and ? wildcard characters *glob patterns* (short for "global patterns").

Glob patterns let you specify patterns of filenames. For example, you could run the dir or ls command to display all the files and folders in the cwd. But if you wanted to see just the Python files, dir *.py or ls *.py would display only the files that end in *.py*. The glob pattern *.py means "any group of characters, followed by .py":

```
C:\Users\Al>dir *.py
 Volume in drive C is Windows
 Volume Serial Number is DFF3-8658

 Directory of C:\Users\Al

03/24/2019  10:45 PM             8,399 conwaygameoflife.py
03/24/2019  11:00 PM             7,896 test1.py
10/29/2019  08:18 PM            21,254 wizcoin.py
               3 File(s)         37,549 bytes
               0 Dir(s)  506,300,776,448 bytes free
```

The glob pattern records201?.txt means "records201, followed by any single character, followed by .txt." This would match record files for the years *records2010.txt* to *records2019.txt* (as well as filenames, such as *records201X.txt*). The glob pattern records20??.txt would match any two characters, such as *records2021.txt* or *records20AB.txt*.

Change Directories with cd

Running cd *[destination folder]* changes the shell's cwd to the destination folder:

```
C:\Users\Al>cd Desktop

C:\Users\Al\Desktop>
```

The shell displays the cwd as part of its prompt, and any folders or files used in commands will be interpreted relative to this directory.

If the folder has spaces in its name, enclose the name in double quotes. To change the cwd to the user's home folder, enter **cd ~** on macOS and Linux, and **cd %USERPROFILE%** on Windows.

On Windows, if you also want to change the current drive, you'll first need to enter the drive name as a separate command:

```
C:\Users\Al>d:

D:\>cd BackupFiles

D:\BackupFiles>
```

To change to the parent directory of the cwd, use the **..** folder name:

```
C:\Users\Al>cd ..

C:\Users>
```

List Folder Contents with dir and ls

On Windows, the dir command displays the folders and files in the cwd. The ls command does the same thing on macOS and Linux. You can display the contents of another folder by running dir *[another folder]* or ls *[another folder]*.

The -l and -a switches are useful arguments for the ls command. By default, ls displays only the names of files and folders. To display a long listing format that includes file size, permissions, last modification timestamps, and other information, use -l. By convention, the macOS and Linux operating systems treat files beginning with a period as configuration files and keep them hidden from normal commands. You can use -a to make ls display all files, including hidden ones. To display both the long listing format and all files, combine the switches as ls -al. Here's an example in a macOS or Linux terminal window:

```
al@ubuntu:~$ ls
Desktop     Downloads         mu_code  Pictures  snap        Videos
Documents   examples.desktop  Music    Public    Templates
al@ubuntu:~$ ls -al
total 112
drwxr-xr-x 18 al    al    4096 Aug  4 18:47 .
drwxr-xr-x  3 root  root  4096 Jun 17 18:11 ..
-rw-------  1 al    al    5157 Aug  2 20:43 .bash_history
-rw-r--r--  1 al    al     220 Jun 17 18:11 .bash_logout
-rw-r--r--  1 al    al    3771 Jun 17 18:11 .bashrc
drwx------ 17 al    al    4096 Jul 30 10:16 .cache
drwx------ 14 al    al    4096 Jun 19 15:04 .config
drwxr-xr-x  2 al    al    4096 Aug  4 17:33 Desktop
--snip--
```

The Windows analog to ls -al is the dir command. Here's an example in a Windows terminal window:

```
C:\Users\Al>dir
 Volume in drive C is Windows
 Volume Serial Number is DFF3-8658

 Directory of C:\Users\Al

06/12/2019  05:18 PM    <DIR>          .
06/12/2019  05:18 PM    <DIR>          ..
12/04/2018  07:16 PM    <DIR>          .android
--snip--
08/31/2018  12:47 AM            14,618 projectz.ipynb
10/29/2014  04:34 PM           121,474 foo.jpg
```

List Subfolder Contents with dir /s and find

On Windows, running dir /s displays the cwd's folders and their subfolders. For example, the following command displays every *.py* file in my *C:\github\ ezgmail* folder and all of its subfolders:

```
C:\github\ezgmail>dir /s *.py
 Volume in drive C is Windows
 Volume Serial Number is DEE0-8982

 Directory of C:\github\ezgmail

06/17/2019  06:58 AM             1,396 setup.py
               1 File(s)          1,396 bytes

 Directory of C:\github\ezgmail\docs

12/07/2018  09:43 PM             5,504 conf.py
               1 File(s)          5,504 bytes

 Directory of C:\github\ezgmail\src\ezgmail

06/23/2019  07:45 PM            23,565 __init__.py
12/07/2018  09:43 PM                56 __main__.py
               2 File(s)         23,621 bytes

     Total Files Listed:
               4 File(s)         30,521 bytes
               0 Dir(s)   505,407,283,200 bytes free
```

The find . -name command does the same thing on macOS and Linux:

```
al@ubuntu:~/Desktop$ find . -name "*.py"
./someSubFolder/eggs.py
./someSubFolder/bacon.py
./spam.py
```

The . tells find to start searching in the cwd. The -name option tells find to find folders and filenames by name. The "*.py" tells find to display folders and files with names that match the *.py pattern. Note that the find command requires the argument after -name to be enclosed in double quotes.

Copy Files and Folders with copy and cp

To create a duplicate of a file or folder in a different directory, run **copy** *[source file or folder] [destination folder]* or **cp** *[source file or folder] [destination folder]*. Here's an example in a Linux terminal window:

```
al@ubuntu:~/someFolder$ ls
hello.py   someSubFolder
al@ubuntu:~/someFolder$ cp hello.py someSubFolder
al@ubuntu:~/someFolder$ cd someSubFolder
al@ubuntu:~/someFolder/someSubFolder$ ls
hello.py
```

Move Files and Folders with move and mv

On Windows, you can move a source file or folder to a destination folder by running move *[source file or folder] [destination folder]*. The mv *[source file or folder] [destination folder]* command does the same thing on macOS and Linux.

Here's an example in a Linux terminal window:

```
al@ubuntu:~/someFolder$ ls
hello.py   someSubFolder
al@ubuntu:~/someFolder$ mv hello.py someSubFolder
al@ubuntu:~/someFolder$ ls
someSubFolder
al@ubuntu:~/someFolder$ cd someSubFolder/
al@ubuntu:~/someFolder/someSubFolder$ ls
hello.py
```

The *hello.py* file has moved from *~/someFolder* to *~/someFolder/someSubFolder* and no longer appears in its original location.

Rename Files and Folders with ren and mv

Running ren *[file or folder] [new name]* renames the file or folder on Windows, and mv *[file or folder] [new name]* does so on macOS and Linux. Note that you can use the mv command on macOS and Linux for moving *and* renaming a file. If you supply the name of an existing folder for the second argument, the mv command moves the file or folder there. If you supply a name that doesn't match an existing file or folder, the mv command renames the file or folder. Here's an example in a Linux terminal window:

```
al@ubuntu:~/someFolder$ ls
hello.py   someSubFolder
```

```
al@ubuntu:~/someFolder$ mv hello.py goodbye.py
al@ubuntu:~/someFolder$ ls
goodbye.py  someSubFolder
```

The *hello.py* file now has the name *goodbye.py*.

Delete Files and Folders with del and rm

To delete a file or folder on Windows, run del *[file or folder]*. To do so on macOS and Linux, run rm *[file]* (rm is short for, remove).

These two delete commands have some slight differences. On Windows, running del on a folder deletes all of its files, but not its subfolders. The del command also won't delete the source folder; you must do so with the rd or rmdir commands, which I'll explain in "Delete Folders with rd and rmdir" on page 34. Additionally, running del *[folder]* won't delete any files inside the subfolders of the source folder. You can delete the files by running del /s /q *[folder]*. The /s runs the del command on the subfolders, and the /q essentially means "be quiet and don't ask me for confirmation." Figure 2-4 illustrates this difference.

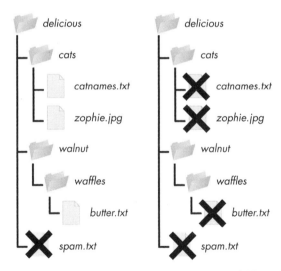

Figure 2-4: The files are deleted in these example folders when you run del delicious *(left) or* del /s /q delicious *(right).*

On macOS and Linux, you can't use the rm command to delete folders. But you can run rm -r *[folder]* to delete a folder and all of its contents. On Windows, rd /s /q *[folder]* will do the same thing. Figure 2-5 illustrates this task.

```
rd /s /q delicious
rm -r delicious
```

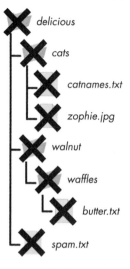

Figure 2-5: The files are deleted in these example folders
when you run rd /s /q delicious or rm -r delicious.

Make Folders with md and mkdir

Running md *[new folder]* creates a new, empty folder on Windows, and running mkdir *[new folder]* does so on macOS and Linux. The mkdir command
also works on Windows, but md is easier to type.

Here's an example in a Linux terminal window:

```
al@ubuntu:~/Desktop$ mkdir yourScripts
al@ubuntu:~/Desktop$ cd yourScripts
❶ al@ubuntu:~/Desktop/yourScripts$ ls
al@ubuntu:~/Desktop/yourScripts$
```

Notice that the newly created *yourScripts* folder is empty; nothing
appears when we run the ls command to list the folder's contents ❶.

Delete Folders with rd and rmdir

Running rd *[source folder]* deletes the source folder on Windows, and rmdir
[source folder] deletes the source folder on macOS and Linux. Like mkdir,
the rmdir command also works on Windows, but rd is easier to type. The
folder must be empty before you can remove it.

Here's an example in a Linux terminal window:

```
al@ubuntu:~/Desktop$ mkdir yourScripts
al@ubuntu:~/Desktop$ ls
yourScripts
```

```
al@ubuntu:~/Desktop$ rmdir yourScripts
al@ubuntu:~/Desktop$ ls
al@ubuntu:~/Desktop$
```

In this example, we created an empty folder named *yourScripts* and then removed it.

To delete nonempty folders (along with all the folders and files it contains), run **rd /s/q** *[source folder]* on Windows or **rm -rf** *[source folder]* on macOS and Linux.

Find Programs with where and which

Running where *[program]* on Windows or which *[program]* on macOS and Linux tells you the exact location of the program. When you enter a command on the command line, your computer checks for the program in the folders listed in the PATH environment variable (although Windows checks the cwd first).

These commands can tell you which executable Python program is run when you enter python in the shell. If you have multiple Python versions installed, your computer might have several executable programs of the same name. The one that is run depends on the order of folders in your PATH environment variable, and the where and which commands will output it:

```
C:\Users\Al>where python
C:\Users\Al\AppData\Local\Programs\Python\Python38\python.exe
```

In this example, the folder name indicates that the Python version run from the shell is located at *C:\Users\Al\AppData\Local\Programs\Python\Python38*.

Clear the Terminal with cls and clear

Running cls on Windows or clear on macOS and Linux will clear all the text in the terminal window. This is useful if you simply want to start with a fresh-looking terminal window.

Environment Variables and PATH

All running processes of a program, no matter the language in which it's written, have a set of variables called *environment variables* that can store a string. Environment variables often hold systemwide settings that every program would find useful. For example, the TEMP environment variable holds the file path where any program can store temporary files. When the operating system runs a program (such as a command line), the newly created process receives its own copy of the operating system's environment variables and values. You can change a process's environment variables independently of the operating system's set of environment variables. But those changes apply only to the process, not to the operating system or any other process.

I discuss environment variables in this chapter because one such variable, PATH, can help you run your programs from the command line.

Viewing Environment Variables

You can see a list of the terminal window's environment variables by running set (on Windows) or env (on macOS and Linux) from the command line:

```
C:\Users\Al>set
ALLUSERSPROFILE=C:\ProgramData
APPDATA=C:\Users\Al\AppData\Roaming
CommonProgramFiles=C:\Program Files\Common Files
--snip--
USERPROFILE=C:\Users\Al
VBOX_MSI_INSTALL_PATH=C:\Program Files\Oracle\VirtualBox\
windir=C:\WINDOWS
```

The text on the left side of the equal sign (=) is the environment variable name, and the text on the right side is the string value. Every process has its own set of environment variables, so different command lines can have different values for their environment variables.

You can also view the value of a single environment variable with the echo command. Run echo %HOMEPATH% on Windows or echo $HOME on macOS and Linux to view the value of the HOMEPATH or HOME environment variables, respectively, which contain the current user's home folder. On Windows, it looks like this:

```
C:\Users\Al>echo %HOMEPATH%
\Users\Al
```

On macOS or Linux, it looks like this:

```
al@al-VirtualBox:~$ echo $HOME
/home/al
```

If that process creates another process (such as when a command line runs the Python interpreter), that child process receives its own copy of the parent process's environment variables. The child process can change the values of its environment variables without affecting the parent process's environment variables, and vice versa.

You can think of the operating system's set of environment variables as the "master copy" from which a process copies its environment variables. The operating system's environment variables change less frequently than a Python program's. In fact, most users never directly touch their environment variable settings.

Working with the PATH Environment Variable

When you enter a command, like python on Windows or python3 on macOS and Linux, the terminal checks for a program with that name in the folder you're currently in. If it doesn't find it there, it will check the folders listed in the PATH environment variable.

For example, on my Windows computer, the *python.exe* program file is located in the *C:\Users\Al\AppData\Local\Programs\Python\Python38* folder. To run it, I have to enter C:\Users\Al\AppData\Local\Programs\Python\Python38\python.exe, or switch to that folder first and then enter python.exe.

This lengthy pathname requires a lot of typing, so instead I add this folder to the PATH environment variable. Then, when I enter python.exe, the command line searches for a program with this name in the folders listed in PATH, saving me from having to type the entire file path.

Because environment variables can contain only a single string value, adding multiple folder names to the PATH environment variable requires using a special format. On Windows, semicolons separate the folder names. You can view the current PATH value with the path command:

```
C:\Users\Al>path
C:\Path;C:\WINDOWS\system32;C:\WINDOWS;C:\WINDOWS\System32\Wbem;
--snip--
C:\Users\Al\AppData\Local\Microsoft\WindowsApps
```

On macOS and Linux, colons separate the folder names:

```
al@ubuntu:~$ echo $PATH
/home/al/.local/bin:/usr/local/sbin:/usr/local/bin:/usr/sbin:/usr/bin:/sbin:/
bin:/usr/games:/usr/local/games:/snap/bin
```

The order of the folder names is important. If I have two files named *someProgram.exe* in *C:\WINDOWS\system32* and *C:\WINDOWS*, entering someProgram.exe will run the program in *C:\WINDOWS\system32* because that folder appears first in the PATH environment variable.

If a program or command you enter doesn't exist in the cwd or any of the directories listed in PATH, the command line will give you an error, such as command not found or not recognized as an internal or external command. If you didn't make a typo, check which folder contains the program and see if it appears in the PATH environment variable.

Changing the Command Line's PATH Environment Variable

You can change the current terminal window's PATH environment variable to include additional folders. The process for adding folders to PATH varies slightly between Windows and macOS/Linux. On Windows, you can run the path command to add a new folder to the current PATH value:

```
❶ C:\Users\Al>path C:\newFolder;%PATH%

❷ C:\Users\Al>path
C:\newFolder;C:\Path;C:\WINDOWS\system32;C:\WINDOWS;C:\WINDOWS\System32\Wbem;
--snip--
C:\Users\Al\AppData\Local\Microsoft\WindowsApps
```

The %PATH% part ❶ expands to the current value of the PATH environment variable, so you're adding the new folder and a semicolon to the beginning of the existing PATH value. You can run the path command again to see the new value of PATH ❷.

On macOS and Linux, you can set the PATH environment variable with syntax similar to an assignment statement in Python:

```
❶ al@al-VirtualBox:~$ PATH=/newFolder:$PATH
❷ al@al-VirtualBox:~$ echo $PATH
/newFolder:/home/al/.local/bin:/usr/local/sbin:/usr/local/bin:/usr/sbin:/usr/
bin:/sbin:/bin:/usr/games:/usr/local/games:/snap/bin
```

The $PATH part ❶ expands to the current value of the PATH environment variable, so you're adding the new folder and a colon to the existing PATH value. You can run the echo $PATH command again to see the new value of PATH ❷.

But the previous two methods for adding folders to PATH apply only to the current terminal window and any programs run from it after the addition. If you open a new terminal window, it won't have your changes. Permanently adding folders requires changing the operating system's set of environment variables.

Permanently Adding Folders to PATH on Windows

Windows has two sets of environment variables: *system environment variables* (which apply to all users) and *user environment variables* (which override the system environment variable but apply to the current user only). To edit them, click the Start menu and then enter **Edit environment variables for your account**, which opens the Environment Variables window, as shown in Figure 2-6.

Select **Path** from the user variable list (not the system variable list), click **Edit**, add the new folder name in the text field that appears (don't forget the semicolon separator), and click **OK**.

This interface isn't the easiest to work with, so if you're frequently editing environment variables on Windows, I recommend installing the free Rapid Environment Editor software from *https://www.rapidee.com/*. Note that after installing it, you must run this software as the administrator to edit system environment variables. Click the Start menu, type **Rapid Environment Editor**, right-click the software's icon, and click **Run as administrator**.

From the Command Prompt, you can permanently modify the system PATH variable using the setx command:

```
C:\Users\Al>setx /M PATH "C:\newFolder;%PATH%"
```

You'll need to run the Command Prompt as the administrator to run the setx command.

Figure 2-6: The Environment Variables window on Windows

Permanently Adding Folders to PATH on macOS and Linux

To add folders to the PATH environment variables for all terminal windows on macOS and Linux, you'll need to modify the *.bashrc* text file in your home folder and add the following line:

```
export PATH=/newFolder:$PATH
```

This line modifies PATH for all future terminal windows. On macOS Catalina and later versions, the default shell program has changed from Bash to Z Shell, so you'll need to modify *.zshrc* in the home folder instead.

Running Python Programs Without the Command Line

You probably already know how to run programs from whatever launcher your operating system provides. Windows has the Start menu, macOS has the Finder and Dock, and Ubuntu Linux has Dash. Programs will add themselves to these launchers when you install them. You can also double-click a program's icon in a file explorer app (such as File Explorer on Windows, Finder on macOS, and Files on Ubuntu Linux) to run them.

But these methods don't apply to your Python programs. Often, double-clicking a *.py* file will open the Python program in an editor or IDE instead of running it. And if you try running Python directly, you'll just open the Python interactive shell. The most common way of running a Python program is opening it in an IDE and clicking the Run menu option or executing it in the command line. Both methods are tedious if you simply want to launch a Python program.

Instead, you can set up your Python programs to easily run them from your operating system's launcher, just like other applications you've installed. The following sections detail how to do this for your particular operating system.

Running Python Programs on Windows

On Windows, you can run Python programs in a few other ways. Instead of opening a terminal window, you can press WIN-R to open the Run dialog and enter `py C:\path\to\yourScript.py`, as shown in Figure 2-7. The *py.exe* program is installed at *C:\Windows\py.exe*, which is already in the PATH environment variable, and the *.exe* file extension is optional when you are running programs.

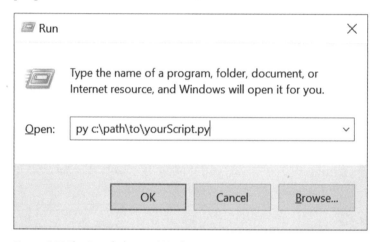

Figure 2-7: The Run dialog on Windows

Still, this method requires you to enter your script's full path. Also, the terminal window that displays the program's output will automatically close when the program ends, and you might miss some output.

You can solve these problems by creating a *batch script*, which is a small text file with the *.bat* file extension that can run multiple terminal commands at once, much like a shell script in macOS and Linux. You can use a text editor, such as Notepad, to create these files. Make a new text file containing the following two lines:

```
@py.exe C:\path\to\yourScript.py %*
@pause
```

Replace this path with the absolute path to your program, and save this file with a *.bat* file extension (for example, *yourScript.bat*). The @ sign at the start of each command prevents it from being displayed in the terminal window, and the %* forwards any command line arguments entered after the batch filename to the Python script. The Python script, in turn, reads the command line arguments in the sys.argv list. This batch file will spare you from having to type the Python program's full absolute path every time you want to run it. The @pause command adds Press any key to continue... to the end of the Python script to prevent the program's window from disappearing too quickly.

I recommend you place all of your batch and *.py* files in a single folder that already exists in the PATH environment variable, such as your home folder at *C:\Users\<USERNAME>*. With a batch file set up, you can run your Python script by simply pressing WIN-R, entering the name of your batch file (entering the *.bat* file extension is optional), and pressing ENTER.

Running Python Programs on macOS

On macOS, you can create a shell script to run your Python scripts by creating a text file with the *.command* file extension. Make one in a text editor, such as TextEdit, and add the following content:

```
#!/usr/bin/env bash
python3 /path/to/yourScript.py
```

Save this file in your home folder. In a terminal window, make this shell script executable by running chmod u+x *yourScript.command*. Now you should be able to click the Spotlight icon (or press COMMAND-SPACE) and enter the name of your shell script to run it. The shell script, in turn, will run your Python script.

Running Python Programs on Ubuntu Linux

There isn't a quick way to run your Python scripts on Ubuntu Linux like there is in Windows and macOS, although you can shorten some of the steps involved. First, make sure your *.py* file is in your home folder. Second, add this line as the first line of your *.py* file:

```
#!/usr/bin/env python3
```

This is called a *shebang line*, and it tells Ubuntu that when you run this file, you want to use python3 to run it. Third, add the execute permission to this file by running the chmod command from the terminal:

```
al@al-VirtualBox:~$ chmod u+x yourScript.py
```

Now whenever you want to quickly run your Python script, you can press CTRL-ALT-T to open a new terminal window. This terminal will be set to the home folder, so you can simply enter ./yourScript.py to run this script. The ./ is required because it tells Ubuntu that *yourScript.py* exists in the cwd (the home folder, in this case).

Summary

Environment setup involves all the steps necessary to get your computer into a state where you can easily run your programs. It requires you to know several low-level concepts about how your computer works, such as the filesystem, file paths, processes, the command line, and environment variables.

The filesystem is how your computer organizes all the files on your computer. A file is a complete, absolute file path or a file path relative to the cwd. You'll navigate the filesystem through the command line. The command line has several other names, such as terminal, shell, and console, but they all refer to the same thing: the text-based program that lets you enter commands. Although the command line and the names of common commands are slightly different between Windows and macOS/Linux, they effectively perform the same tasks.

When you enter a command or program name, the command line checks the folders listed in the PATH environment variable for the name. This is important to understand to figure out any command not found errors you might encounter. The steps for adding new folders to the PATH environment variable are also slightly different between Windows and macOS/Linux.

Becoming comfortable with the command line takes time because there are so many commands and command line arguments to learn. Don't worry if you spend a lot of time searching for help online; this is what experienced software developers do every day.

PART 2

BEST PRACTICES, TOOLS, AND TECHNIQUES

3

CODE FORMATTING WITH BLACK

Code formatting is applying a set of rules
to source code to give it a certain appear-
ance. Although unimportant to the computer parsing your program, code formatting is
vital for readability, which is necessary for maintaining your code. If your code is difficult for humans
(whether it's you or a co-worker) to understand, it will
be hard to fix bugs or add new features. Formatting
code isn't a mere cosmetic issue. Python's readability
is a critical reason for the language's popularity.

This chapter introduces you to Black, a code formatting tool that can
automatically format your source code into a consistent, readable style
without changing your program's behavior. Black is useful, because it's
tedious to manually format your code in a text editor or IDE. You'll first
learn about the rationalization for the code style choices Black makes.
Then you'll learn how to install, use, and customize the tool.

How to Lose Friends and Alienate Co-Workers

We can write code in many ways that result in identical behavior. For example, we can write a list with a single space after each comma and use one kind of quote character consistently:

```
spam = ['dog', 'cat', 'moose']
```

But even if we wrote the list with a varying number of spaces and quote styles, we'd still have syntactically valid Python code:

```
spam= [ 'dog'  ,'cat',"moose"]
```

Programmers who prefer the former approach might like the visual separation that the spaces add and the uniformity of the quote characters. But programmers sometimes choose the latter, because they don't want to worry about details that have no impact on whether the program works correctly.

Beginners often ignore code formatting because they're focused on programming concepts and language syntax. But it's valuable for beginners to establish good code formatting habits. Programming is difficult enough, and writing understandable code for others (or for yourself in the future) can minimize this problem.

Although you might start out coding on your own, programming is often a collaborative activity. If several programmers working on the same source code files write in their own style, the code can become an inconsistent mess, even if it runs without error. Or worse, the programmers will constantly be reformatting each other's code to their own style, wasting time and causing arguments. Deciding whether to, say, put one or zero spaces after a comma is a matter of personal preference. These style choices can be much like deciding which side of the road to drive on; it doesn't matter whether people drive on the right side of the road or the left side, as long as everyone consistently drives on the same side.

Style Guides and PEP 8

An easy way to write readable code is to follow a *style guide*, a document that outlines a set of formatting rules a software project should follow. The *Python Enhancement Proposal 8* (*PEP 8*) is one such style guide written by the Python core development team. But some software companies have established their own style guides as well.

You can find PEP 8 online at *https://www.python.org/dev/peps/pep-0008/*. Many Python programmers view PEP 8 as an authoritative set of rules, although the PEP 8 creators argue otherwise. The "A Foolish Consistency Is the Hobgoblin of Little Minds" section of the guide reminds the reader that maintaining consistency and readability within a project, rather than adhering to any individual formatting rule, is the prime reason for enforcing style guides.

PEP 8 even includes the following advice: "Know when to be inconsistent—sometimes style guide recommendations just aren't applicable. When in doubt, use your best judgment." Whether you follow all of it, some of it, or none of it, it's worthwhile to read the PEP 8 document.

Because we're using the Black code formatter, our code will follow Black's style guide, which is adapted from PEP 8's style guide. You should learn these code formatting guidelines, because you might not always have Black conveniently at hand. The Python code guidelines you learn in this chapter also generally apply to other languages, which might not have automatic formatters available.

I don't like everything about how Black formats code, but I take that as the sign of a good compromise. Black uses formatting rules that programmers can live with, letting us spend less time arguing and more time programming.

Horizontal Spacing

Empty space is just as important for readability as the code you write. These spaces help separate distinct parts of code from each other, making them easier to identify. This section explains *horizontal spacing*—that is, the placement of blank space within a single line of code, including the indentation at the front of the line.

Use Space Characters for Indentation

Indentation is the whitespace at the beginning of a code line. You can use one of two whitespace characters, a space or a tab, to indent your code. Although either character works, the best practice is to use spaces instead of tabs for indentation.

The reason is that these two characters behave differently. A space character is always rendered on the screen as a string value with a single space, like this ' '. But a tab character, which is rendered as a string value containing an escape character, or '\t', is more ambiguous. Tabs often, but not always, render as a variable amount of spacing so the following text begins at the next tab stop. The tab stop are positioned every eight spaces across the width of a text file. You can see this variation in the following interactive shell example, which first separates words with space characters and then with tab characters:

```
>>> print('Hello there, friend!\nHow are you?')
Hello there, friend!
How are you?
>>> print('Hello\tthere,\tfriend!\nHow\tare\tyou?')
Hello   there,  friend!
How     are     you?
```

Because tabs represent a varying width of whitespace, you should avoid using them in your source code. Most code editors and IDEs will automatically insert four or eight space characters when you press the TAB key instead of one tab character.

You also can't use tabs *and* spaces for indentation in the same block of code. Using both for indentation was such a source of tedious bugs in earlier Python programs that Python 3 won't even run code indented like this; it raises a `TabError: inconsistent use of tabs and spaces in indentation` exception instead. Black automatically converts any tab characters you use for indentation into four space characters.

As for the length of each indentation level, the common practice in Python code is four spaces per level of indentation. The space characters in the following example have been marked with periods to make them visible:

```
def getCatAmount():
....numCats = input('How many cats do you have?')
....if int(numCats) < 6:
........print('You should get more cats.')
```

The four-space standard has practical benefits compared to the alternatives; using eight spaces per level of indentation results in code that quickly runs up against line length limits, whereas using two space characters per level of indentation can make the differences in indentation hard to see. Programmers often don't consider other amounts, such as three or six spaces, because they, and binary computing in general, have a bias for numbers that are powers of two: 2, 4, 8, 16, and so on.

Spacing Within a Line

Horizontal spacing has more to it than just the indentation. Spaces are important for making different parts of a code line appear visually distinct. If you never use space characters, your line can end up dense and hard to parse. The following subsections provide some spacing rules to follow.

Put a Single Space Between Operators and Identifiers

If you don't leave spaces between operators and identifiers, your code will appear to run together. For example, this line has spaces separating operators and variables:

```
YES: blanks = blanks[:i] + secretWord[i] + blanks[i + 1 :]
```

This line removes all spacing:

```
NO:   blanks=blanks[:i]+secretWord[i]+blanks[i+1:]
```

In both cases, the code uses the + operator to add three values, but without spacing, the + in `blanks[i+1:]` can appear to be adding a fourth value. The spaces make it more obvious that this + is part of a slice for the value in `blanks`.

Put No Spaces Before Separators and a Single Space After Separators

We separate the items lists and dictionaries, as well as the parameters in function def statements, using comma characters. You should place no spaces before these commas and a single space after them, as in this example:

```
YES: def spam(eggs, bacon, ham):
YES:    weights = [42.0, 3.1415, 2.718]
```

Otherwise, you'll end up with "bunched up" code that is harder to read:

```
NO:  def spam(eggs,bacon,ham):
NO:     weights = [42.0,3.1415,2.718]
```

Don't add spaces before the separator, because that unnecessarily draws the eye to the separator character:

```
NO:  def spam(eggs , bacon , ham):
NO:     weights = [42.0 , 3.1415 , 2.718]
```

Black automatically inserts a space after commas and removes spaces before them.

Don't Put Spaces Before or After Periods

Python allows you to insert spaces before and after the periods marking the beginning of a Python attribute, but you should avoid doing so. By not placing spaces there, you emphasize the connection between the object and its attribute, as in this example:

```
YES: 'Hello, world'.upper()
```

If you put spaces before or after the period, the object and attribute look like they're unrelated to each other:

```
NO:  'Hello, world' . upper()
```

Black automatically removes spaces surrounding periods.

Don't Put Spaces After a Function, Method, or Container Name

We can readily identify function and method names because they're followed by a set of parentheses, so don't put a space between the name and the opening parenthesis. We would normally write a function call like this:

```
YES: print('Hello, world!')
```

But adding a space makes this singular function call look like it's two separate things:

```
NO:  print ('Hello, world!')
```

Black removes any spaces between a function or method name and its opening parenthesis.

Similarly, don't put spaces before the opening square bracket for an index, slice, or key. We normally access items inside a container type (such as a list, dictionary, or tuple) without adding spaces between the variable name and opening square bracket, like this:

```
YES: spam[2]
YES: spam[0:3]
YES: pet['name']
```

Adding a space once again makes the code look like two separate things:

```
NO:   spam [2]
NO:   spam    [0:3]
NO:   pet ['name']
```

Black removes any spaces between the variable name and opening square bracket.

Don't Put Spaces After Opening Brackets or Before Closing Brackets

There should be no spaces separating parentheses, square brackets, or braces and their contents. For example, the parameters in a def statement or values in a list should start and end immediately after and before their parentheses and square brackets:

```
YES: def spam(eggs, bacon, ham):
YES:     weights = [42.0, 3.1415, 2.718]
```

You should not put a space after an opening or before a closing parentheses or square brackets:

```
NO:   def spam( eggs, bacon, ham ):
NO:       weights = [ 42.0, 3.1415, 2.718 ]
```

Adding these spaces doesn't improve the code's readability, so it's unnecessary. Black removes these spaces if they exist in your code.

Put Two Spaces Before End-of-Line Comments

If you add comments to the end of a code line, put two spaces after the end of the code and before the # character that begins the comment:

```
YES: print('Hello, world!')  # Display a greeting.
```

The two spaces make it easier to distinguish the code from the comment. A single space, or worse, no space, makes it more difficult to notice this separation:

```
NO:   print('Hello, world!') # Display a greeting.
NO:   print('Hello, world!')# Display a greeting.
```

Black puts two spaces between the end of the code and the start of the comment.

In general, I advise against putting comments at the end of a code line, because they can make the line too lengthy to read onscreen.

Vertical Spacing

Vertical spacing is the placement of blank lines between lines of code. Just as a new paragraph in a book keeps sentences from forming a wall of text, vertical spacing can group certain lines of code together and separate those groups from one another.

PEP 8 has several guidelines for inserting blank lines in code: it states that you should separate functions with two blank lines, classes with two blank lines, and methods within a class with one blank line. Black automatically follows these rules by inserting or removing blank lines in your code, turning this code:

```
NO:  class ExampleClass:
         def exampleMethod1():
             pass
         def exampleMethod2():
             pass
     def exampleFunction():
         pass
```

. . . into this code:

```
YES: class ExampleClass:
         def exampleMethod1():
             pass

         def exampleMethod2():
             pass


     def exampleFunction():
         pass
```

A Vertical Spacing Example

What Black *can't* do is decide where blank lines within your functions, methods, or global scope should go. Which of those lines to group together is a subjective decision that is up to the programmer.

For example, let's look at the `EmailValidator` class in *validators.py* in the Django web app framework. It's not necessary for you to understand how

this code works. But pay attention to how blank lines separate the __call__()
method's code into four groups:

```
--snip--
    def __call__(self, value):
    ❶ if not value or '@' not in value:
            raise ValidationError(self.message, code=self.code)

    ❷ user_part, domain_part = value.rsplit('@', 1)

    ❸ if not self.user_regex.match(user_part):
            raise ValidationError(self.message, code=self.code)

    ❹ if (domain_part not in self.domain_whitelist and
                not self.validate_domain_part(domain_part)):
            # Try for possible IDN domain-part
            try:
                domain_part = punycode(domain_part)
            except UnicodeError:
                pass
            else:
                if self.validate_domain_part(domain_part):
                    return
            raise ValidationError(self.message, code=self.code)
--snip--
```

Even though there are no comments to describe this part of the code, the
blank lines indicate that the groups are conceptually distinct from each other.
The first group ❶ checks for an @ symbol in the value parameter. This task
is different from that of the second group ❷, which splits the email address
string in value into two new variables, user_part and domain_part. The third ❸
and fourth ❹ groups use these variables to validate the user and domain
parts of the email address, respectively.

Although the fourth group has 11 lines, far more than the other groups,
they're all related to the same task of validating the domain of the email
address. If you felt that this task was really composed of multiple subtasks,
you could insert blank lines to separate them.

The programmer for this part of Django decided that the domain validation lines should all belong to one group, but other programmers might
disagree. Because it's subjective, Black won't modify the vertical spacing
within functions or methods.

Vertical Spacing Best Practices

One of Python's lesser-known features is that you can use a semicolon to
separate multiple statements on a single line. This means that the following
two lines:

```
print('What is your name?')
name = input()
```

. . . can be written on the same line if separated by a semicolon:

```
print('What is your name?'); name = input()
```

As you do when using commas, you should put no space before the semicolon and one space after it.

For statements that end with a colon, such as `if`, `while`, `for`, `def`, or `class` statements, a single-line block, like the call to `print()` in this example:

```
if name == 'Alice':
    print('Hello, Alice!')
```

. . . can be written on the same line as its `if` statement:

```
if name == 'Alice': print('Hello, Alice!')
```

But just because Python allows you to include multiple statements on the same line doesn't make it a good practice. It results in overly wide lines of code and too much content to read on a single line. Black splits these statements into separate lines.

Similarly, you can import multiple modules with a single `import` statement:

```
import math, os, sys
```

Even so, PEP 8 recommends that you split this statement into one `import` statement per module:

```
import math
import os
import sys
```

If you write separate lines for imports, you'll have an easier time spotting any additions or removals of imported modules when you're comparing changes in a version control system's diff tool. (Version control systems, such as Git, are covered in Chapter 12.)

PEP 8 also recommends grouping `import` statements into the following three groups in this order:

1. Modules in the Python standard library, like `math`, `os`, and `sys`
2. Third-party modules, like Selenium, Requests, or Django
3. Local modules that are a part of the program

These guidelines are optional, and Black won't change the formatting of your code's `import` statements.

Black: The Uncompromising Code Formatter

Black automatically formats the code inside your *.py* files. Although you should understand the formatting rules covered in this chapter, Black can do all the actual styling for you. If you're working on a coding project with others, you can instantly settle many arguments on how to format code by just letting Black decide.

You can't change many of the rules that Black follows, which is why it's described as "the uncompromising code formatter." Indeed, the tool's name comes from Henry Ford's quote about the automobile colors choices he offered his customers: "You can have any color you want, as long as it's black."

I've just described the exact styles that Black uses; you can find Black's full style guide at *https://black.readthedocs.io/en/stable/the_black_code_style.html*.

Installing Black

Install Black using the `pip` tool that comes with Python. In Windows, do this by opening a Command Prompt window and entering the following:

```
C:\Users\Al\>python -m pip install --user black
```

On macOS and Linux, open a Terminal window and enter **python3** rather than `python` (you should do this for all the instructions in this book that use `python`):

```
Als-MacBook-Pro:~ al$ python3 -m pip install --user black
```

The -m option tells Python to run the `pip` module as an application, which some Python modules are set up to do. Test that the installation was successful by running **python -m black**. You should see the message No paths given. Nothing to do. rather than No module named black.

Running Black from the Command Line

You can run Black for any Python file from the Command Prompt or Terminal window. In addition, your IDE or code editor can run Black in the background. You'll find instructions for getting Black to work with Jupyter Notebook, Visual Studio Code, PyCharm, and other editors on Black's home page at *https://github.com/psf/black/*.

Let's say that you want to format a file called *yourScript.py* automatically. From the command line in Windows, run the following (on macOS and Linux, use the `python3` command instead of `python`):

```
C:\Users\Al>python -m black yourScript.py
```

After you run this command, the content of *yourScript.py* will be formatted according to Black's style guide.

Your PATH environment variable might already be set up to run Black directly, in which case you can format *yourScript.py* by simply entering the following:

```
C:\Users\Al>black yourScript.py
```

If you want to run Black over every *.py* file in a folder, specify a single folder instead of an individual file. The following Windows example formats every file in the *C:\yourPythonFiles* folder, including its subfolders:

```
C:\Users\Al>python -m black C:\yourPythonFiles
```

Specifying the folder is useful if your project contains several Python files and you don't want to enter a command for each one.

Although Black is fairly strict about how it formats code, the next three subsections describe a few options that you *can* change. To see the full range of options that Black offers, run `python -m black --help`.

Adjusting Black's Line Length Setting

The standard line of Python code is 80 characters long. The history of the 80 character line dates back to the era of punch card computing in the 1920s when IBM introduced punch cards that had 80 columns and 12 rows. The 80 column standard remained for the printers, monitors, and command line windows developed over the next several decades.

But in the 21st century, high-resolution screens can display text that is more than 80 characters wide. A longer line length can keep you from having to scroll vertically to view a file. A shorter line length can keep too much code from crowding on a single line and allow you to compare two source code files side by side without having to scroll horizontally.

Black uses a default of 88 characters per line for the rather arbitrary reason that it is 10 percent more than the standard 80 character line. My preference is to use 120 characters. To tell Black to format your code with, for example, a 120-character line length limit, use the `-l 120` (that's the lowercase letter *L,* not the number 1) command line option. On Windows, the command looks like this:

```
C:\Users\Al>python -m black -l 120 yourScript.py
```

No matter what line length limit you choose for your project, all *.py* files in a project should use the same limit.

Disabling Black's Double-Quoted Strings Setting

Black automatically changes any string literals in your code from using single quotes to double quotes unless the string contains double quote characters, in which case it uses single quotes. For example, let's say *yourScript.py* contains the following:

```
a = 'Hello'
b = "Hello"
c = 'Al\'s cat, Zophie.'
d = 'Zophie said, "Meow"'
e = "Zophie said, \"Meow\""
f = '''Hello'''
```

Running Black on *yourScript.py* would format it like this:

```
❶ a = "Hello"
   b = "Hello"
   c = "Al's cat, Zophie."
❷ d = 'Zophie said, "Meow"'
   e = 'Zophie said, "Meow"'
❸ f = """Hello"""
```

Black's preference for double quotes makes your Python code look similar to code written in other programming languages, which often use double quotes for string literals. Notice that the strings for variables a, b, and c use double quotes. The string for variable d retains its original single quotes to avoid escaping any double quotes within the string ❷. Note that Black also uses double quotes for Python's triple-quoted, multiline strings ❸.

But if you want Black to leave your string literals as you wrote them and not change the type of quotes used, pass it the -S command line option. (Note that the *S* is uppercase.) For example, running Black on the original *yourScript.py* file in Windows would produce the following output:

```
C:\Users\Al>python -m black -S yourScript.py
All done!
1 file left unchanged.
```

You can also use the -l line length limit and -S options in the same command:

```
C:\Users\Al>python -m black -l 120 -S yourScript.py
```

Previewing the Changes Black Will Make

Although Black won't rename your variable or change your program's behavior, you might not like the style changes Black makes. If you want to stick to your original formatting, you could either use version control for your source code or maintain your own backups. Alternatively, you can preview the changes Black would make without letting it actually alter your files by running Black with the --diff command line option. In Windows, it looks like this:

```
C:\Users\Al>python -m black --diff yourScript.py
```

This command outputs the differences in the diff format commonly used by version control software, but it's generally readable by humans. For example, if *yourScript.py* contains the line weights=[42.0,3.1415,2.718], running the --diff option would display this result:

```
C:\Users\Al\>python -m black --diff yourScript.py
--- yourScript.py       2020-12-07 02:04:23.141417 +0000
```

```
+++ yourScript.py          2020-12-07 02:08:13.893578 +0000
@@ -1 +1,2 @@
-weights=[42.0,3.1415,2.718]
+weights = [42.0, 3.1415, 2.718]
```

The minus sign indicates that Black would remove the line `weights=[42.0,3.1415,2.718]` and replace it with the line prefixed with a plus sign: `weights = [42.0, 3.1415, 2.718]`. Keep in mind that once you've run Black to change your source code files, there's no way to undo this change. You need to either make backup copies of your source code or use version control software, such as Git, before running Black.

Disabling Black for Parts of Your Code

As great as Black is, you might not want it to format some sections of your code. For example, I like to do my own special spacing whenever I'm lining up multiple related assignment statements, as in the following example:

```
# Set up constants for different time amounts:
SECONDS_PER_MINUTE = 60
SECONDS_PER_HOUR    = 60 * SECONDS_PER_MINUTE
SECONDS_PER_DAY     = 24 * SECONDS_PER_HOUR
SECONDS_PER_WEEK    = 7  * SECONDS_PER_DAY
```

Black would remove the additional spaces before the = assignment operator, making them, in my opinion, less readable:

```
# Set up constants for different time amounts:
SECONDS_PER_MINUTE = 60
SECONDS_PER_HOUR = 60 * SECONDS_PER_MINUTE
SECONDS_PER_DAY = 24 * SECONDS_PER_HOUR
SECONDS_PER_WEEK = 7 * SECONDS_PER_DAY
```

By adding `# fmt: off` and `# fmt: on` comments, we can tell Black to turn off its code formatting for these lines and then resume code formatting afterward:

```
# Set up constants for different time amounts:
# fmt: off
SECONDS_PER_MINUTE = 60
SECONDS_PER_HOUR    = 60 * SECONDS_PER_MINUTE
SECONDS_PER_DAY     = 24 * SECONDS_PER_HOUR
SECONDS_PER_WEEK    = 7  * SECONDS_PER_DAY
# fmt: on
```

Running Black on this file now won't affect the unique spacing, or any other formatting, in the code between these two comments.

Summary

Although good formatting can be invisible, poor formatting can make reading code frustrating. Style is subjective, but the software development field generally agrees on what constitutes good and poor formatting while still leaving room for personal preferences.

Python's syntax makes it rather flexible when it comes to style. If you're writing code that nobody else will ever see, you can write it however you like. But much of software development is collaborative. Whether you're working with others on a project or simply want to ask more experienced developers to review your work, formatting your code to fit accepted style guides is important.

Formatting your code in an editor is a boring task that you can automate with a tool like Black. This chapter covered several of the guidelines that Black follows to make your code more readable, including spacing code vertically and horizontally, which keeps it from being too dense to read easily, and setting a limit on how long each line should be. Black enforces these rules for you, preventing potential style arguments with collaborators.

But there's more to code style than spacing and deciding between single and double quotes. For instance, choosing descriptive variable names is also a critical factor for code readability. Although automated tools like Black can make *syntactic* decisions, such as the amount of spacing code should have, they can't make *semantic* decisions, such as what a good variable name is. That responsibility is yours, and we'll discuss this topic in the next chapter.

4

CHOOSING UNDERSTANDABLE NAMES

"The two hardest problems in computer science are naming things, cache invalidation, and off-by-one errors." This classic joke, attributed to Leon Bambrick and based on a quote by Phil Karlton, contains a kernel of truth: it's hard to come up with good names, formally called *identifiers*, for variables, functions, classes, and anything else in programming. Concise, descriptive names are important for your program's readability. But creating names is easier said than done. If you were moving to a new house, labeling all your moving boxes as "Stuff" would be concise but not descriptive. A descriptive name for a programming book might be *Invent Your Own Computer Games with Python*, but it's not concise.

Unless you're writing "throwaway" code that you don't intend to maintain after you run the program once, you should put some thought into selecting good names in your program. If you simply use a, b, and c for variable names, your future self will expend unnecessary effort to remember what these variables were initially used for.

Names are a subjective choice that you must make. An automated formatting tool, such as Black, described in Chapter 3, can't decide what you should name your variables. This chapter provides you with some guidelines to help you choose suitable names and avoid poor names. As always, these guidelines aren't written in stone: use your judgment to decide when to apply them to your code.

METASYNTACTIC VARIABLES

We commonly use a *metasyntactic variable* in tutorials or code snippets when we need a generic variable name. In Python, we often name variables spam, eggs, bacon, and ham in code examples where the variable name isn't important. That's why this book uses these names in code examples; they aren't meant for you to use them in real-world programs. These names come from Monty Python's "Spam" sketch (*https://en.wikipedia.org/wiki/Spam_(Monty_Python)*). The names foo and bar are also common for metasyntactic variables. These are derived from FUBAR, the World War II era US Army slang acronym that indicates a situation is "[messed] up beyond all recognition."

Casing Styles

Because Python identifiers are case sensitive and cannot contain whitespace, programmers use several styles for identifiers that include multiple words:

snake_case separates words with an underscore, which looks like a flat snake in between each word. This case often implies that all letters are lowercase, although constants are often written in UPPER_SNAKE_CASE.

camelCase separates words by capitalizing the start of each word after the first. This case often implies the first word begins with a lowercase letter. The uppercase letters look like a camel's humps.

PascalCase, named for its use in the Pascal programming language, is similar to camelCase but capitalizes the first word as well.

Casing is a code formatting issue and we cover it in Chapter 3. The most common styles are snake_case and camelCase. Either is fine to use as long as your project consistently uses one or the other, not both.

PEP 8's Naming Conventions

The PEP 8 document introduced in Chapter 3 has some recommendations for Python naming conventions:

- All letters should be ASCII letters—that is, uppercase and lowercase English letters that don't have accent marks.
- Modules should have short, all lowercase names.
- Class names should be written in PascalCase.
- Constant variables should be written in uppercase SNAKE_CASE.
- Function, method, and variable names should be written in lowercase snake_case.
- The first argument for methods should always be named self in lowercase.
- The first argument for class methods should always be named cls in lowercase.
- Private attributes in classes should always begin with an underscore (_).
- Public attributes in classes should never begin with an underscore (_).

You can bend or break these rules as required. For example, although English is the dominant language in programming, you can use letter characters in any language as identifiers: コンピューター = 'laptop' is syntactically valid Python code. As you can see in this book, my preference for variable names goes against PEP 8, because I use camelCase rather than snake_case. PEP 8 contains a reminder that a programmer doesn't need to strictly follow PEP 8. The important readability factor isn't which style you choose but consistency in using that style.

You can read PEP 8's "Naming Conventions" section online at *https://www.python.org/dev/peps/pep-0008/#naming-conventions*.

Appropriate Name Length

Obviously, names shouldn't be too long or too short. Long variable names are tedious to type, whereas short variable names can be confusing or mysterious. Because code is read more often than it's written, it's safer to err on the side of too long variable names. Let's look at some examples of names that are too short and too long.

Too Short Names

The most common naming mistake is choosing names that are too short. Short names often make sense to you when you first write them, but their precise meaning can be lost a few days or weeks later. Let's consider a few types of short names.

- A *one- or two-letter name* like g probably refers to some other word that begins with g, but there are many such words. Acronyms and names that are only one or two letters long are easy for you to write but difficult for someone else to read. This also applies to . . .

- . . . an *abbreviated name* like mon, which could stand for *monitor, month, monster,* or any number of words.

- A *single-word name* like start can be vague: the start of what? Such names could be missing context that isn't readily apparent when read by someone else.

One- or two-letter, abbreviated, or single-word names might be understandable to you, but you always need to keep in mind that other programmers (or even you a few weeks into the future) will have difficulty understanding their meaning.

There are some exceptions where short variable names are fine. For example, it's common to use i (for *index*) as a variable name in for loops that loop over a range of numbers or indexes of a list, and j and k (because they come after *i* in the alphabet) if you have nested loops:

```
>>> for i in range(10):
...     for j in range(3):
...         print(i, j)
...
0 0
0 1
0 2
1 0
--snip--
```

Another exception is using x and y for Cartesian coordinates. In most other cases, I caution against using single-letter variable names. Although it might be tempting to use, say, w and h as shorthand for width and height, or n as shorthand for number, these meanings might not be apparent to others.

DN'T DRP LTTRS FRM YR SRC CD

Don't drop letters from your source code. Although dropped letters in names like memcpy (memory copy) and strcmp (string compare) were popular in the C programming language before the 1990s, they're an unreadable style of naming that you shouldn't use today. If a name isn't easily pronounceable, it isn't easily understood.

Additionally, feel free to use short phrases that can make your code read like plain English. For example, number_of_trials is more readable than simply number_trials.

Too Long Names

In general, the larger the name's scope, the more descriptive it should be. A short name like payment is fine for a local variable inside a single, short function. But payment might not be descriptive enough if you use it for a global variable across a 10,000-line program, because such a large program might process multiple kinds of payment data. A more descriptive name, such as salesClientMonthlyPayment or annual_electric_bill_payment, could be more suitable. The additional words in the name provide more context and resolve ambiguity.

It's better to be overly descriptive than not descriptive enough. But there are guidelines for determining when longer names are unnecessary.

Prefixes in Names

The use of common prefixes in names could indicate unnecessary detail in the name. If a variable is an attribute of a class, the prefix might provide information that doesn't need to be in the variable name. For example, if you have a Cat class with a weight attribute, it's obvious that weight refers to the cat's weight. So the name catWeight would be overly descriptive and unnecessarily long.

Similarly, an old and now obsolete practice is the use of *Hungarian notation,* the practice of including an abbreviation of the data type in names. For example, the name strName indicates that the variable contains a string value, and iVacationDays indicates that the variable contains an integer. Modern languages and IDEs can relay this data type information to the programmer without the need for these prefixes, making Hungarian notation an unnecessary practice today. If you find you're including the name of a data type in your names, consider removing it.

On the other hand, the is and has prefixes for variables that contain Boolean values, or functions and methods that return Boolean values, make those names more readable. Consider the following use of a variable named is_vehicle and a method named has_key():

```
if item_under_repair.has_key('tires'):
  is_vehicle = True
```

The has_key() method and is_vehicle variable support a plain English reading of the code: "if the item under repair has a key named 'tires,' then it's true that the item is a vehicle."

Similarly, adding units to your names can provide useful information. A weight variable that stores a floating-point value is ambiguous: is the weight in pounds, kilograms, or tons? This unit information isn't a data type, so including a prefix or suffix of kg or lbs or tons isn't the same as Hungarian notation. If you aren't using a weight-specific data type that contains unit information, naming the variable something like weight_kg could be prudent. Indeed, in 1999 the Mars Climate Orbiter robotic space probe was lost when software supplied by Lockheed Martin produced calculations in imperial standard units, whereas NASA's systems used metric, resulting in an incorrect trajectory. The spacecraft reportedly cost $125 million.

Sequential Numeric Suffixes in Names

Sequential numeric suffixes in your names indicate that you might need to change the variable's data type or add different details to the name. Numbers alone often don't provide enough information to distinguish these names.

Variable names like `payment1`, `payment2`, and `payment3` don't tell the person reading the code what the difference is between these payment values. The programmer should probably refactor these three variables into a single list or tuple variable named `payments` that contains three values.

Functions with calls like `makePayment1(amount)`, `makePayment2(amount)`, and so on should probably be refactored into a single function that accepts an integer argument: `makePayment(1, amount)`, `makePayment(2, amount)`, and so on. If these functions have different behaviors that justify separate functions, the meaning behind the numbers should be stated in the name: `makeLowPriorityPayment` `(amount)` and `makeHighPriorityPayment(amount)`, or `make1stQuarterPayment(amount)` and `make2ndQuarterPayment(amount)`, for example.

If you have a valid reason for choosing names with sequential numeric suffixes, it's fine to use them. But if you're using these names because it's an easy choice to make, consider revising them.

Make Names Searchable

For all but the smallest programs, you'll probably need to use your editor or IDE's CTRL-F "find" feature to locate where your variables and functions are referenced. If you choose a short, generic variable name, such as `num` or `a`, you'll end up with several false matches. To make the name easy to find immediately, form unique names by using longer variable names that contain specific details.

Some IDEs will have refactoring features that can identify names based on how your program uses them. For example, a common feature is a "rename" tool that can differentiate between variables named `num` and `number`, as well as between local `num` and global `num` variables. But you should still choose names as though these tools weren't available.

Keeping this rule in mind will naturally help you pick descriptive names instead of generic ones. The name `email` is vague, so consider a more descriptive name like `emailAddress`, `downloadEmailAttachment`, `emailMessage`, or `replyToAddress`. Not only would such a name be more precise, it would be easier to find in your source code files as well.

Avoid Jokes, Puns, and Cultural References

At one of my previous software jobs, our codebase contained a function named `gooseDownload()`. I had no idea what this meant, because the product we were creating had nothing to do with birds or the downloading of birds. When I found the more-senior co-worker who had originally written the function, he explained that *goose* was meant as a verb, as in "goose the engine." I had no idea what this phrase meant, either. He had to further explain that "goose the engine" was automotive jargon that meant press down on the gas pedal to make the engine go faster. Thus, `gooseDownload()` was a function to make downloads go faster. I nodded my head and went back to my desk. Years later, after this co-worker left the company, I renamed his function to `increaseDownloadSpeed()`.

When choosing names in your program, you might be tempted to use jokes, puns, or cultural references to add some levity to your code. Don't do this. Jokes can be hard to convey in text, and the joke probably won't be as funny in the future. Puns can also be easy to miss, and handling repeat bug reports from co-workers who confused a pun for a typo can be quite punishing.

Culture-specific references can get in the way of communicating your code's intent clearly. The internet makes it easier than ever to share source code with strangers around the world who won't necessarily be fluent in English or understand English jokes. As noted earlier in the chapter, the names `spam`, `eggs`, and `bacon` used in Python documentation reference a Monty Python comedy sketch, but we use these as metasyntactic variables only; it's inadvisable to use them in real-world code.

The best policy is to write your code in a way that non-native English speakers can readily understand: polite, direct, and humorless. My former co-worker might have thought `gooseDownload()` was a funny joke, but nothing kills a joke faster than having to explain it.

Don't Overwrite Built-in Names

You should also never use Python's built-in names for your own variables. For example, if you name a variable `list` or `set`, you'll overwrite Python's `list()` and `set()` functions, possibly causing bugs later in your code. The `list()` function creates list objects. But overwriting it can lead to this error:

```
>>> list(range(5))
[0, 1, 2, 3, 4]
❶ >>> list = ['cat', 'dog', 'moose']
❷ >>> list(range(5))
Traceback (most recent call last):
  File "<stdin>", line 1, in <module>
TypeError: 'list' object is not callable
```

If we assign a list value to the name `list` ❶, we'll lose the original `list()` function. Attempting to call `list()` ❷ would result in a `TypeError`. To find out whether Python is already using a name, type it into the interactive shell or try to import it. You'll get a `NameError` or `ModuleNotFoundError` if the name isn't being used. For example, Python uses the names open and test but doesn't use spam and eggs:

```
>>> open
<built-in function open >
>>> import test
>>> spam
Traceback (most recent call last):
  File "<stdin>", line 1, in <module>
NameError: name 'spam' is not defined
>>> import eggs
Traceback (most recent call last):
  File "<stdin>", line 1, in <module>
ModuleNotFoundError: No module named 'eggs'
```

Some commonly overwritten Python names are all, any, date, email, file, format, hash, id, input, list, min, max, object, open, random, set, str, sum, test, and type. Don't use these names for your identifiers.

Another common problem is naming your *.py* files the same names as third-party modules. For example, if you installed the third-party Pyperclip module but also created a *pyperclip.py* file, an import pyperclip statement imports *pyperclip.py* instead of the Pyperclip module. When you try to call Pyperclip's copy() or paste() functions, you'll get an error saying they don't exist:

```
>>> # Run this code with a file named pyperclip.py in the current folder.
>>> import pyperclip # This imports your pyperclip.py, not the real one.
>>> pyperclip.copy('hello')
Traceback (most recent call last):
  File "<stdin>", line 1, in <module>
AttributeError: module 'pyperclip' has no attribute 'copy'
```

Be aware of overwriting existing names in your Python code, especially if you're unexpectedly getting these has no attribute error messages.

The Worst Possible Variable Names Ever

The name data is a terrible, generic variable name, because literally all variables contain data. The same goes for naming variables var, which is a bit like naming your pet dog "Dog." The name temp is common for variables that temporarily hold data but is still a poor choice: after all, from a Zen perspective, all variables are temporary. Unfortunately, these names occur frequently despite their vagueness; avoid using them in your code.

If you need a variable to hold the statistical variance of your temperature data, please use the name temperatureVariance. It should go without saying that the name tempVarData would be a poor choice.

Summary

Choosing names has nothing to do with algorithms or computer science, and yet it's a vital part of writing readable code. Ultimately, the names you use in your code are up to you, but be aware of the many guidelines that exist. The PEP 8 document recommends several naming conventions, such as lowercase names for modules and PascalCase names for classes. Names shouldn't be too short or too long. But it's often better to err on the side of too descriptive instead of not detailed enough.

A name should be concise but descriptive. A name that is easy to find using a CTRL-F search feature is the sign of a distinct and descriptive variable. Think about how searchable your name is to determine whether you're using a too generic name. Also, consider whether a programmer who doesn't speak fluent English would understand the name: avoid using jokes, puns, and cultural references in your names; instead, choose names that are polite, direct, and humorless.

Although many of the suggestions in this chapter are simply guidelines, you should always avoid names Python's standard library already uses, such as `all`, `any`, `date`, `email`, `file`, `format`, `hash`, `id`, `input`, `list`, `min`, `max`, `object`, `open`, `random`, `set`, `str`, `sum`, `test`, and `type`. Using these names could cause subtle bugs in your code.

The computer doesn't care whether your names are descriptive or vague. Names make code easier to read by humans, not easier to run by computers. If your code is readable, it's easy to understand. If it's easy to understand, it's easy to change. And if it's easy to change, it's easier to fix bugs or add new features. Using understandable names is a foundational step to producing quality software.

5

FINDING CODE SMELLS

Code that causes a program to crash is obviously wrong, but crashes aren't the only indicator of issues in your programs. Other signs can suggest the presence of more subtle bugs or unreadable code. Just as the smell of gas can indicate a gas leak or the smell of smoke could indicate a fire, a *code smell* is a source code pattern that signals potential bugs. A code smell doesn't necessarily mean a problem exists, but it does mean you should investigate your program.

This chapter lists several common code smells. It takes much less time and effort to prevent a bug than to encounter, understand, and fix a bug later. Every programmer has stories of spending hours debugging only to find that the fix involved changing a single line of code. For this reason, even a whiff of a potential bug should give you pause, prompting you to double-check that you aren't creating future problems.

Of course, a code smell isn't necessarily a problem. Ultimately, whether to address or ignore a code smell is a judgment call for you to make.

Duplicate Code

The most common code smell is *duplicate code*. Duplicate code is any source code that you could have created by copying and pasting some other code into your program. For example, this short program contains duplicate code. Notice that it asks how the user is feeling three times:

```
print('Good morning!')
print('How are you feeling?')
feeling = input()
print('I am happy to hear that you are feeling ' + feeling + '.')
print('Good afternoon!')
print('How are you feeling?')
feeling = input()
print('I am happy to hear that you are feeling ' + feeling + '.')
print('Good evening!')
print('How are you feeling?')
feeling = input()
print('I am happy to hear that you are feeling ' + feeling + '.')
```

Duplicate code is a problem because it makes changing the code difficult; a change you make to one copy of the duplicate code must be made to every copy of it in the program. If you forget to make a change somewhere, or if you make different changes to different copies, your program will likely end up with bugs.

The solution to duplicate code is to *deduplicate* it; that is, make it appear once in your program by placing the code in a function or loop. In the following example, I've moved the duplicate code into a function and then repeatedly called that function:

```
def askFeeling():
    print('How are you feeling?')
    feeling = input()
    print('I am happy to hear that you are feeling ' + feeling + '.')

print('Good morning!')
askFeeling()
print('Good afternoon!')
askFeeling()
print('Good evening!')
askFeeling()
```

In this next example, I've moved the duplicate code into a loop:

```
for timeOfDay in ['morning', 'afternoon', 'evening']:
    print('Good ' + timeOfDay + '!')
    print('How are you feeling?')
    feeling = input()
    print('I am happy to hear that you are feeling ' + feeling + '.')
```

You could also combine these two techniques and use a function and a loop:

```
def askFeeling(timeOfDay):
    print('Good ' + timeOfDay + '!')
    print('How are you feeling?')
    feeling = input()
    print('I am happy to hear that you are feeling ' + feeling + '.')

for timeOfDay in ['morning', 'afternoon', 'evening']:
    askFeeling(timeOfDay)
```

Notice that the code that produces the "Good morning/afternoon/evening!" messages is similar but not identical. In the third improvement to the program, I parameterized the code to deduplicate the identical parts. Meanwhile, the timeOfDay parameter and timeOfDay loop variable replace the parts that differ. Now that I've deduplicated this code by removing the extra copies, I only need to make any necessary changes in one place.

As with all code smells, avoiding duplicate code isn't a hard-and-fast rule you must always follow. In general, the longer the duplicate code section or the more duplicate copies that appear in your program, the stronger the case for deduplicating it. I don't mind copying and pasting code once or even twice. But I generally start to consider deduplicating code when three or four copies exist in my program.

Sometimes, code is just not worth the trouble of deduplicating. Compare the first code example in this section to the most recent one. Although the duplicate code is longer, it's simple and straightforward. The deduplicated example does the same thing but involves a loop, a new timeOfDay loop variable, and a new function with a parameter that is also named timeOfDay.

Duplicate code is a code smell because it makes your code harder to change consistently. If several duplicates are in your program, the solution is to place code inside a function or loop so it appears only once.

Magic Numbers

It's no surprise that programming involves numbers. But some of the numbers that appear in your source code can confuse other programmers (or you a couple weeks after writing them). For example, consider the number 604800 in the following line:

```
expiration = time.time() + 604800
```

The `time.time()` function returns an integer representing the current time. We can assume that the expiration variable will represent some point 604,800 seconds into the future. But 604800 is rather mysterious: what's the significance of this expiration date? A comment can help clarify:

```
expiration = time.time() + 604800  # Expire in one week.
```

This is a good solution, but an even better one is to replace these "magic" numbers with constants. *Constants* are variables whose names are written in uppercase letters to indicate that their values shouldn't change after their initial assignment. Usually, constants are defined as global variables at the top of the source code file:

```
# Set up constants for different time amounts:
SECONDS_PER_MINUTE = 60
SECONDS_PER_HOUR   = 60 * SECONDS_PER_MINUTE
SECONDS_PER_DAY    = 24 * SECONDS_PER_HOUR
SECONDS_PER_WEEK   = 7  * SECONDS_PER_DAY

--snip--

expiration = time.time() + SECONDS_PER_WEEK  # Expire in one week.
```

You should use separate constants for magic numbers that serve different purposes, even if the magic number is the same. For example, there are 52 cards in a deck of playing cards and 52 weeks in a year. But if you have both amounts in your program, you should do something like the following:

```
NUM_CARDS_IN_DECK = 52
NUM_WEEKS_IN_YEAR = 52

print('This deck contains', NUM_CARDS_IN_DECK, 'cards.')
print('The 2-year contract lasts for', 2 * NUM_WEEKS_IN_YEAR, 'weeks.')
```

When you run this code, the output will look like this:

```
This deck contains 52 cards.
The 2-year contract lasts for 104 weeks.
```

Using separate constants allows you to change them independently in the future. Note that constant variables should never change values while the program is running. But this doesn't mean that the programmer can never update them in the source code. For example, if a future version of the code includes a joker card, you can change the cards constant without affecting the weeks one:

```
NUM_CARDS_IN_DECK = 53
NUM_WEEKS_IN_YEAR = 52
```

The term *magic number* can also apply to non-numeric values. For example, you might use string values as constants. Consider the following program, which asks the user to enter a direction and displays a warning if the direction is north. A 'nrth' typo causes a bug that prevents the program from displaying the warning:

```
while True:
    print('Set solar panel direction:')
    direction = input().lower()
    if direction in ('north', 'south', 'east', 'west'):
        break

print('Solar panel heading set to:', direction)
❶ if direction == 'nrth':
    print('Warning: Facing north is inefficient for this panel.')
```

This bug can be hard to detect: the typo in the 'nrth' string ❶ is still syntactically correct Python. The program doesn't crash, and it's easy to overlook the lack of a warning message. But if we used constants and made this same typo, the typo would cause the program to crash because Python would notice that a NRTH constant doesn't exist:

```
# Set up constants for each cardinal direction:
NORTH = 'north'
SOUTH = 'south'
EAST = 'east'
WEST = 'west'

while True:
    print('Set solar panel direction:')
    direction = input().lower()
    if direction in (NORTH, SOUTH, EAST, WEST):
        break

print('Solar panel heading set to:', direction)
❶ if direction == NRTH:
    print('Warning: Facing north is inefficient for this panel.')
```

The NameError exception raised by the code line with the NRTH typo ❶ makes the bug immediately obvious when you run this program:

```
Set solar panel direction:
west
Solar panel heading set to: west
Traceback (most recent call last):
  File "panelset.py", line 14, in <module>
    if direction == NRTH:
NameError: name 'NRTH' is not defined
```

Magic numbers are a code smell because they don't convey their purpose, making your code less readable, harder to update, and prone to undetectable typos. The solution is to use constant variables instead.

Commented-Out Code and Dead Code

Commenting out code so it doesn't run is fine as a temporary measure. You might want to skip some lines to test other functionality, and commenting them out makes them easy to add back in later. But if commented-out code remains in place, it's a complete mystery why it was removed and under what condition it might ever be needed again. Consider the following example:

```
doSomething()
#doAnotherThing()
doSomeImportantTask()
doAnotherThing()
```

This code prompts many unanswered questions: Why was doAnother Thing() commented out? Will we ever include it again? Why wasn't the second call to doAnotherThing() commented out? Were there originally two calls to doAnotherThing(), or was there one call that was moved after doSome ImportantTask()? Is there a reason we shouldn't remove the commented-out code? There are no readily available answers to these questions.

Dead code is code that is unreachable or logically can never run. For example, code inside a function but after a return statement, code in an if statement block with an always False condition, or code in a function that is never called is all dead code. To see this in practice, enter the following into the interactive shell:

```
>>> import random
>>> def coinFlip():
...     if random.randint(0, 1):
...         return 'Heads!'
...     else:
...         return 'Tails!'
...     return 'The coin landed on its edge!'
...
>>> print(coinFlip())
Tails!
```

The return 'The coin landed on its edge!' line is dead code because the code in the if and else blocks returns before the execution could ever reach that line. Dead code is misleading because programmers reading it assume that it's an active part of the program when it's effectively the same as commented-out code.

Stubs are an exception to these code smell rules. These are placeholders for future code, such as functions or classes that have yet to be implemented. In lieu of real code, a stub contains a pass statement, which does nothing. (It's also called a *no operation* or *no-op*.) The pass statement exists so you can create stubs in places where the language syntax requires some code:

```
>>> def exampleFunction():
...     pass
...
```

When this function is called, it does nothing. Instead, it indicates that code will eventually be added in.

Alternatively, to avoid accidentally calling an unimplemented function, you can stub it with a `raise NotImplementedError` statement. This will immediately indicate that the function isn't yet ready to be called:

```
>>> def exampleFunction():
...     raise NotImplementedError
...
>>> exampleFunction()
Traceback (most recent call last):
  File "<stdin>", line 1, in <module>
  File "<stdin>", line 2, in exampleFunction
NotImplementedError
```

Raising a `NotImplementedError` will warn you whenever your program calls a stub function or method by accident.

Commented-out code and dead code are code smells because they can mislead programmers into thinking that the code is an executable part of the program. Instead, remove them and use a version control system, such as Git or Subversion, to keep track of changes. Version control is covered in Chapter 12. With version control, you can remove the code from your program and, if needed, easily add it back in later.

Print Debugging

Print debugging is the practice of placing temporary `print()` calls in a program to display the values of variables and then rerunning the program. The process often follows these steps:

1. Notice a bug in your program.
2. Add `print()` calls for some variables to find out what they contain.
3. Rerun the program.
4. Add some more `print()` calls because the earlier ones didn't show enough information.
5. Rerun the program.
6. Repeat the previous two steps a few more times before finally figuring out the bug.
7. Rerun the program.
8. Realize you forgot to remove some of the `print()` calls and remove them.

Print debugging is deceptively quick and simple. But it often requires multiple iterations of rerunning the program before you display the information you need to fix your bug. The solution is to use a debugger or set up logfiles for your program. By using a debugger, you can run your programs one line of code at a time and inspect any variable. Using a debugger might seem slower than simply inserting a `print()` call, but it saves you time in the long run.

Logfiles can record large amounts of information from your program so you can compare one run of it to previous runs. In Python, the built-in logging module provides the functionality you need to easily create logfiles by using just three lines of code:

```
import logging
logging.basicConfig(filename='log_filename.txt', level=logging.DEBUG,
format='%(asctime)s - %(levelname)s - %(message)s')
logging.debug('This is a log message.')
```

After importing the logging module and setting up its basic configuration, you can call logging.debug() to write information to a text file, as opposed to using print() to display it onscreen. Unlike print debugging, calling logging.debug() makes it obvious what output is debugging information and what output is the result of the program's normal run. You can find more information about debugging in Chapter 11 of *Automate the Boring Stuff with Python*, 2nd edition (No Starch, 2019), which you can read online at *https://autbor.com/2e/c11/*.

Variables with Numeric Suffixes

When writing programs, you might need multiple variables that store the same kind of data. In those cases, you might be tempted to reuse a variable name by adding a numeric suffix to it. For example, if you're handling a signup form that asks users to enter their password twice to prevent typos, you might store those password strings in variables named password1 and password2. These numeric suffixes aren't good descriptions of what the variables contain or the differences between them. They also don't indicate how many of these variables there are: is there a password3 or a password4 as well? Try to create distinct names rather than lazily adding numeric suffixes. A better set of names for this password example would be password and confirm_password.

Let's look at another example: if you have a function that deals with start and destination coordinates, you might have the parameters x1, y1, x2, and y2. But the numeric suffix names don't convey as much information as the names start_x, start_y, end_x, and end_y. It's also clearer that the start_x and start_y variables are related to each other, compared to x1 and y1.

If your numeric suffixes extend past 2, you might want to use a list or set data structure to store your data as a collection. For example, you could store the values of pet1Name, pet2Name, pet3Name, and so on in a list called petNames.

This code smell doesn't apply to every variable that simply ends with a number. For example, it's perfectly fine to have a variable named enableIPv6, because "6" is part of the "IPv6" proper name, not a numeric suffix. But if you're using numeric suffixes for a series of variables, consider replacing them with a data structure, such as a list or dictionary.

Classes That Should Just Be Functions or Modules

Programmers who use languages such as Java are used to creating classes to organize their program's code. For example, let's look at this example Dice class, which has a roll() method:

```
>>> import random
>>> class Dice:
...     def __init__(self, sides=6):
...         self.sides = sides
...     def roll(self):
...         return random.randint(1, self.sides)
...
>>> d = Dice()
>>> print('You rolled a', d.roll())
You rolled a 1
```

This might seem like well-organized code, but think about what our actual needs are: a random number between 1 and 6. We could replace this entire class with a simple function call:

```
>>> print('You rolled a', random.randint(1, 6))
You rolled a 6
```

Compared to other languages, Python uses a casual approach to organizing code, because its code isn't required to exist in a class or other boilerplate structure. If you find you're creating objects simply to make a single function call, or if you're writing classes that contain only static methods, these are code smells that indicate you might be better off writing functions instead.

In Python, we use modules rather than classes to group functions together. Because classes must be in a module anyway, putting this code in classes just adds an unnecessary layer of organization to your code. Chapters 15 through 17 discuss these object-oriented design principles in more detail. Jack Diederich's PyCon 2012 talk "Stop Writing Classes" covers other ways that you might be overcomplicating your Python code.

List Comprehensions Within List Comprehensions

List comprehensions are a concise way to create complex list values. For example, to create a list of strings of digits for the numbers 0 through 100, excluding all multiples of 5, you'd typically need a for loop:

```
>>> spam = []
>>> for number in range(100):
...     if number % 5 != 0:
...         spam.append(str(number))
...
>>> spam
['1', '2', '3', '4', '6', '7', '8', '9', '11', '12', '13', '14', '16', '17',
--snip--
'86', '87', '88', '89', '91', '92', '93', '94', '96', '97', '98', '99']
```

Alternatively, you can create this same list in a single line of code by using the list comprehension syntax:

```
>>> spam = [str(number) for number in range(100) if number % 5 != 0]
>>> spam
['1', '2', '3', '4', '6', '7', '8', '9', '11', '12', '13', '14', '16', '17',
--snip--
'86', '87', '88', '89', '91', '92', '93', '94', '96', '97', '98', '99']
```

Python also has syntax for set comprehensions and dictionary comprehensions:

```
❶ >>> spam = {str(number) for number in range(100) if number % 5 != 0}
>>> spam
{'39', '31', '96', '76', '91', '11', '71', '24', '2', '1', '22', '14', '62',
--snip--
'4', '57', '49', '51', '9', '63', '78', '93', '6', '86', '92', '64', '37'}
❷ >>> spam = {str(number): number for number in range(100) if number % 5 != 0}
>>> spam
{'1': 1, '2': 2, '3': 3, '4': 4, '6': 6, '7': 7, '8': 8, '9': 9, '11': 11,
--snip--
'92': 92, '93': 93, '94': 94, '96': 96, '97': 97, '98': 98, '99': 99}
```

A set comprehension ❶ uses braces instead of square brackets and produces a set value. A dictionary comprehension ❷ produces a dictionary value and uses a colon to separate the key and value in the comprehension.

These comprehensions are concise and can make your code more readable. But notice that the comprehensions produce a list, set, or dictionary based on an iterable object (in this example, the range object returned by the range(100) call). Lists, sets, and dictionaries are iterable objects, which means you could have comprehensions nested inside of comprehensions, as in the following example:

```
>>> nestedIntList = [[0, 1, 2, 3], [4], [5, 6], [7, 8, 9]]
>>> nestedStrList = [[str(i) for i in sublist] for sublist in nestedIntList]
>>> nestedStrList
[['0', '1', '2', '3'], ['4'], ['5', '6'], ['7', '8', '9']]
```

But nested list comprehensions (or nested set and dictionary comprehensions) cram a lot of complexity into a small amount of code, making your code hard to read. It's better to expand the list comprehension into one or more for loops instead:

```
>>> nestedIntList = [[0, 1, 2, 3], [4], [5, 6], [7, 8, 9]]
>>> nestedStrList = []
>>> for sublist in nestedIntList:
...     nestedStrList.append([str(i) for i in sublist])
...
>>> nestedStrList
[['0', '1', '2', '3'], ['4'], ['5', '6'], ['7', '8', '9']]
```

Comprehensions can also contain multiple for expressions, although this tends to produce unreadable code as well. For example, the following list comprehension produces a flattened list from a nested list:

```
>>> nestedList = [[0, 1, 2, 3], [4], [5, 6], [7, 8, 9]]
>>> flatList = [num for sublist in nestedList for num in sublist]
>>> flatList
[0, 1, 2, 3, 4, 5, 6, 7, 8, 9]
```

This list comprehension contains two for expressions, but it's difficult for even experienced Python developers to understand. The expanded form, which uses two for loops, creates the same flattened list but is much easier to read:

```
>>> nestedList = [[0, 1, 2, 3], [4], [5, 6], [7, 8, 9]]
>>> flatList = []
>>> for sublist in nestedList:
...     for num in sublist:
...         flatList.append(num)
...
>>> flatList
[0, 1, 2, 3, 4, 5, 6, 7, 8, 9]
```

Comprehensions are syntactic shortcuts that can produce concise code, but don't go overboard and nest them within each other.

Empty except Blocks and Poor Error Messages

Catching exceptions is one of the primary ways to ensure that your programs will continue to function even when problems arise. When an exception is raised but there is no except block to handle it, the Python program crashes by immediately stopping. This could result in losing your unsaved work or leaving files in a half-finished state.

You can prevent crashes by supplying an except block that contains code for handling the error. But it can be difficult to decide how to handle an error, and programmers can be tempted to simply leave the except block blank with a pass statement. For example, in the following code we use pass to create an except block that does nothing:

```
>>> try:
...     num = input('Enter a number: ')
...     num = int(num)
... except ValueError:
...     pass
...
Enter a number: forty two
>>> num
'forty two'
```

This code doesn't crash when `'forty two'` is passed to `int()` because the `ValueError` that `int()` raises is handled by the except statement. But doing nothing in response to an error might be worse than a crash. Programs crash so they don't continue to run with bad data or in incomplete states, which could lead to even worse bugs later on. Our code doesn't crash when nondigit characters are entered. But now the `num` variable contains a string instead of an integer, which could cause issues whenever the `num` variable gets used. Our except statement isn't handling errors so much as hiding them.

Handling exceptions with poor error messages is another code smell. Look at this example:

```
>>> try:
...     num = input('Enter a number: ')
...     num = int(num)
... except ValueError:
...     print('An incorrect value was passed to int()')
...
Enter a number: forty two
An incorrect value was passed to int()
```

This code doesn't crash, which is good, but it doesn't give the user enough information to know how to fix the problem. Error messages are meant to be read by users, not programmers. Not only does this error message have technical details that a user wouldn't understand, such as a reference to the `int()` function, but it doesn't tell the user how to fix the problem. Error messages should explain what happened as well as what the user should do about it.

It's easier for programmers to quickly write a single, unhelpful description of what happened rather than detailed steps that the user can take to fix the problem. But keep in mind that if your program doesn't handle all possible exceptions that could be raised, it's an unfinished program.

Code Smell Myths

Some code smells aren't really code smells at all. Programming is full of half-remembered bits of bad advice that are taken out of context or stick around long after they've outlived their usefulness. I blame tech book authors who try to pass off their subjective opinions as best practices.

You might have been told some of these practices are code smells, but they're mostly fine. I call them *code smell myths*: they're warnings that you can and should ignore. Let's look at a few of them.

Myth: Functions Should Have Only One return Statement at the End

The "one entry, one exit" idea comes from misinterpreted advice from the days of programming in assembly and FORTRAN languages. These languages allowed you to enter a subroutine (a structure similar to a function) at any point, including in the middle of it, making it hard to debug which parts of the subroutine had been executed. Functions don't have this problem (execution always begins at the start of the function). But the advice lingered, becoming "functions and methods should only have one return statement, which should be at the end of the function or method."

Trying to achieve a single return statement per function or method often requires a convoluted series of if-else statements that's far more confusing than having multiple return statements. Having more than one return statement in a function or method is fine.

Myth: Functions Should Have at Most One try Statement

"Functions and methods should do one thing" is good advice in general. But taking this to mean that exception handling should occur in a separate function goes too far. For example, let's look at a function that indicates whether a file we want to delete is already nonexistent:

```
>>> import os
>>> def deleteWithConfirmation(filename):
...     try:
...         if (input('Delete ' + filename + ', are you sure? Y/N') == 'Y'):
...             os.unlink(filename)
...     except FileNotFoundError:
...         print('That file already did not exist.')
...
```

Proponents of this code smell myth argue that because functions should always do just one thing, and error handling is one thing, we should split this function into two functions. They argue that if you use a try-except statement, it should be the first statement in a function and envelop all of the function's code to look like this:

```
>>> import os
>>> def handleErrorForDeleteWithConfirmation(filename):
...     try:
...         _deleteWithConfirmation(filename)
...     except FileNotFoundError:
...         print('That file already did not exist.')
...
>>> def _deleteWithConfirmation(filename):
...     if (input('Delete ' + filename + ', are you sure? Y/N') == 'Y'):
...         os.unlink(filename)
...
```

This is unnecessarily complicated code. The _deleteWithConfirmation() function is now marked as private with the _ underscore prefix to clarify that it should never be called directly, only indirectly through a call to handleErrorForDeleteWithConfirmation(). This new function's name is awkward, because we call it intending to delete a file, not handle an error to delete a file.

Your functions should be small and simple, but this doesn't mean they should always be limited to doing "one thing" (however you define that). It's fine if your functions have more than one try-except statement and the statements don't envelop all of the function's code.

Myth: Flag Arguments Are Bad

Boolean arguments to function or method calls are sometimes referred to as *flag arguments*. In programming, a *flag* is a value that indicates a binary setting, such as "enabled" or "disabled," and it's often represented by a Boolean value. We can describe these settings as set (that is, True) or cleared (that is, False).

The false belief that flag arguments to function calls are bad is based on the claim that, depending on the flag value, the function does two entirely different things, such as in the following example:

```
def someFunction(flagArgument):
    if flagArgument:
        # Run some code...
    else:
        # Run some completely different code...
```

Indeed, if your function looks like this, you should create two separate functions rather than making an argument decide which half of the function's code to run. But most functions with flag arguments don't do this. For example, you can pass a Boolean value for the sorted() function's reverse keyword argument to determine the sort order. This code wouldn't be improved by splitting the function into two functions named sorted() and reverseSorted() (while also duplicating the amount of documentation required). So the idea that flag arguments are always bad is a code smell myth.

Myth: Global Variables Are Bad

Functions and methods are like mini-programs within your program: they contain code, including local variables that are forgotten when the function returns. This is similar to how a program's variables are forgotten after it terminates. Functions are isolated: their code either performs correctly or has a bug depending on the arguments passed when they're called.

But functions and methods that use global variables lose some of this helpful isolation. Every global variable you use in a function effectively becomes another input to the function, just like the arguments. More arguments mean more complexity, which in turn means a higher likelihood for bugs. If a bug manifests in a function due to a bad value in a global variable, that bad value could have been set anywhere in the program. To search for a likely cause of this bad value, you can't just analyze the code in the function or the line of code calling the function; you must look at the entire program's code. For this reason, you should limit your use of global variables.

For example, let's look at the `calculateSlicesPerGuest()` function in a fictional *partyPlanner.py* program that is thousands of lines long. I've included line numbers to give you a sense of the program's size:

```
1504. def calculateSlicesPerGuest(numberOfCakeSlices):
1505.     global numberOfPartyGuests
1506.     return numberOfCakeSlices / numberOfPartyGuests
```

Let's say when we run this program, we encounter the following exception:

```
Traceback (most recent call last):
  File "partyPlanner.py", line 1898, in <module>
    print(calculateSlicesPerGuest(42))
  File "partyPlanner.py", line 1506, in calculateSlicesPerGuest
    return numberOfCakeSlices / numberOfPartyGuests
ZeroDivisionError: division by zero
```

The program has a zero divide error, caused by the line return `numberOfCakeSlices / numberOfPartyGuests`. The `numberOfPartyGuests` variable must be set to 0 to have caused this, but where did `numberOfPartyGuests` get assigned this value? Because it's a global variable, it could have happened anywhere in the thousands of lines in this program! From the traceback information, we know that `calculateSlicesPerGuest()` was called on line 1898 of our fictional program. If we looked at line 1898, we could find out what argument was passed for the `numberOfCakeSlices` parameter. But the `numberOfPartyGuests` global variable could have been set at any time before the function call.

Note that global constants aren't considered poor programming practice. Because their values never change, they don't introduce complexity into the code the way other global variables do. When programmers mention that "global variables are bad," they aren't referring to constant variables.

Global variables broaden the amount of debugging needed to find where an exception-causing value could have been set. This makes abundant use of global variables a bad idea. But the idea that *all* global variables are bad is a code smell myth. Global variables can be useful in smaller programs or for keeping track of settings that apply across the entire program. If you can avoid using a global variable, that's a sign you probably should avoid using one. But "global variables are bad" is an oversimplified opinion.

Myth: Comments Are Unnecessary

Bad comments are indeed worse than no comments at all. A comment with outdated or misleading information creates more work for the programmer instead of a better understanding. But this potential problem is sometimes used to proclaim that *all* comments are bad. The argument asserts that every comment should be replaced with more readable code, to the point that programs should have no comments at all.

Comments are written in English (or whatever language the programmer speaks), allowing them to convey information to a degree that variable, function, and class names cannot. But writing concise, effective comments is hard. Comments, like code, require rewrites and multiple iterations to get right. We understand the code we write immediately after writing it, so writing comments seems like pointless extra work. As a result, programmers are primed to accept the "comments are unnecessary" viewpoint.

The far more common experience is that programs have too few or no comments rather than too many or misleading comments. Rejecting comments is like saying, "Flying across the Atlantic Ocean in a passenger jet is only 99.999991 percent safe, so I'm going to swim across it instead."

Chapter 10 has more information about how to write effective comments.

Summary

Code smells indicate that there might be a better way to write your code. They don't necessarily require a change, but they should make you take another look. The most common code smell is duplicate code, which can signal an opportunity to place code inside a function or loop. This ensures future code changes will need to be made in only one place. Other code smells include magic numbers, which are unexplained values in your code that can be replaced by constants with descriptive names. Similarly, commented-out code and dead code are never run by the computer, and might mislead programmers who later read the program's code. It's best to remove them and rely on a source control system like Git if you later need to add them back to your program.

Print debugging uses `print()` calls to display debugging information. Although this approach to debugging is easy, it's often faster in the long run to rely on a debugger and logs to diagnose bugs.

Variables with numeric suffixes, such as x1, x2, x3, and so on, are often best replaced with a single variable containing a list. Unlike in languages such as Java, in Python we use modules rather than classes to group functions together. A class that contains a single method or only static methods is a code smell suggesting that you should put that code into a module rather than a class. And although list comprehensions are a concise way to create list values, nested list comprehensions are usually unreadable.

Additionally, any exceptions handled with empty except blocks are a code smell that you're simply silencing the error rather than handling it. A short, cryptic error message is just as useless to the user as no error message.

Along with these code smells are the code smell myths: programming advice that is no longer valid or has, over time, proven counterproductive. These include putting only a single return statement or try-except block in each function, never using flag arguments or global variables, and believing that comments are unnecessary.

Of course, as with all programming advice, the code smells described in this chapter might or might not apply to your project or personal preferences. A best practice isn't an objective measure. As you gain more experience, you'll come to different conclusions about what code is readable or reliable, but the recommendations in this chapter outline issues to consider.

6

WRITING PYTHONIC CODE

Powerful is a meaningless adjective for programming languages. Every programming language describes itself as powerful: the official Python Tutorial begins with the sentence "Python is an easy to learn, powerful programming language." But there's no algorithm that one language can do that another can't, and no unit of measurement to quantify a programming language's "power" (although you certainly can measure the volume at which programmers argue for their favorite language).

But every language does have its own design patterns and gotchas that make up its strengths and weaknesses. To write Python code like a true *Pythonista*, you'll need to know more than just the syntax and standard library. The next step is to learn its *idioms*, or Python-specific coding practices. Certain Python language features lend themselves to writing code in ways that have become known as *pythonic*.

In this chapter, I'll provide several common ways of writing idiomatic Python code along with their unpythonic counterparts. What counts as pythonic can vary from programmer to programmer, but it commonly includes the examples and practices I discuss here. Experienced Python programmers use these techniques, so becoming familiar with them allows you to recognize them in real-world code.

The Zen of Python

The Zen of Python by Tim Peters is a set of 20 guidelines for the design of the Python language and for Python programs. Your Python code doesn't necessarily have to follow these guidelines, but they're good to keep in mind. The Zen of Python is also an *Easter egg*, or hidden joke, that appears when you run import this:

```
>>> import this
The Zen of Python, by Tim Peters

Beautiful is better than ugly.
Explicit is better than implicit.
--snip--
```

NOTE   *Mysteriously, only 19 of the guidelines are written down. Guido van Rossum, creator of Python, reportedly said that the missing 20th aphorism is "some bizarre Tim Peters in-joke." Tim left it blank for Guido to fill, which he seems to never have gotten around to doing.*

In the end, these guidelines are opinions that programmers can argue for or against. Like all good sets of moral codes, they contradict themselves to provide the most flexibility. Here's my interpretation of these aphorisms:

Beautiful is better than ugly.   Beautiful code can be thought of as easy to read and understand. Programmers often write code quickly without concern for readability. The computer will run unreadable code, but unreadable code is difficult for human programmers to maintain and debug. Beauty is subjective, but code that is written without regard for how understandable it is often ugly to others. The reason for Python's popularity is that its syntax isn't cluttered with cryptic punctuation marks like other languages, making it easy to work with.

Explicit is better than implicit. If I'd written only "This is self-explanatory," I would have provided a terrible explanation for this aphorism. Similarly, in code, it's best to be verbose and explicit. You should avoid hiding code's functionality behind obscure language features that require deep language familiarity to fully understand.

Simple is better than complex. Complex is better than complicated. These two aphorisms remind us that we can build anything with simple or complex techniques. If you have a simple problem that requires a shovel, it's overkill to use a 50-ton hydraulic bulldozer. But for an enormous job, the complexity of operating a single bulldozer is preferable to the complications of coordinating a team of 100 shovelers. Prefer simplicity to complexity, but know the limits of simplicity.

Flat is better than nested. Programmers love to organize their code into categories, especially categories that contain subcategories that contain other sub-subcategories. These hierarchies often don't add organization so much as they add bureaucracy. It's okay to write code in just one top-level module or data structure. If your code looks like `spam.eggs.bacon.ham()` or `spam['eggs']['bacon']['ham']`, you're making your code too complicated.

Sparse is better than dense. Programmers often like to cram as much functionality into as little code as possible, as in the following line: `print('\n'.join("%i bytes = %i bits which has %i possiblevalues." % (j, j*8, 256**j-1) for j in (1 << i for i in range(8))))`. Although code like this might impress their friends, it'll infuriate their co-workers who have to try to understand it. Don't make your code try to do too much at once. Code that is spread out over multiple lines is often easier to read than dense one-liners. This aphorism is roughly the same as simple is better than complex.

Readability counts. Although `strcmp()` might obviously mean the "string compare" function to someone who has been programming in C since the 1970s, modern computers have enough memory to write out the full function name. Don't drop letters from your names or write overly terse code. Take the time to come up with descriptive, specific names for your variables and functions. A blank line in between sections of your code can serve the same function as paragraph breaks in a book, letting the reader know which parts are meant to be read together. This aphorism is roughly the same as beautiful is better than ugly.

Special cases aren't special enough to break the rules. Although practicality beats purity. These two aphorisms contradict each other. Programming is full of "best practices" that programmers should strive for in their code. Skirting these practices for a quick hack might be tempting but can lead to a rat's nest of inconsistent, unreadable code. On the other hand, bending over backward to adhere to rules can result in highly abstract, unreadable code. For example, the Java programming language's attempt to fit all code to its object-oriented paradigm often results in lots of boilerplate code for even the smallest program. Walking the line between these two aphorisms becomes easier with experience. In time, you'll learn not only the rules but also when to break them.

Errors should never pass silently. Unless explicitly silenced.   Just because programmers often ignore error messages doesn't mean the program should stop emitting them. Silent errors can happen when functions return error codes or None instead of raising exceptions. These two aphorisms tell us that it's better for a program to fail fast and crash than to silence the error and continue running. The bugs that inevitably happen later on will be harder to debug because they're detected long after the original cause. Although you can always decide to explicitly ignore the errors your programs cause, be sure you're making the conscious choice to do so.

In the face of ambiguity, refuse the temptation to guess.   Computers have made humans superstitious: to exorcise the demons in our computers, we perform the sacred ritual of turning them off and then on. Supposedly this will fix any mysterious problem. But computers are not magic. If your code isn't working, there's a reason why, and only careful, critical thinking will solve the problem. Refuse the temptation to blindly try solutions until something seems to work; often, you've merely masked the problem rather than solved it.

There should be one—and preferably only one—obvious way to do it.   This is a broadside against the Perl programming language's motto, "There's more than one way to do it!" It turns out that having three or four different ways to write code that does the same task is a double-edged sword: you have flexibility in how you write code, but now you have to learn every possible way it could have been written to read other people's code. This flexibility isn't worth the increased effort needed to learn a programming language.

Although that way may not be obvious at first unless you're Dutch. This line is a joke. Guido van Rossum, the creator of Python, is Dutch.

Now is better than never. Although never is often better than *right* now.   These two aphorisms tell us that code that runs slowly is obviously worse than code that runs quickly. But it's better to have to wait for your program to finish than to finish it too early with incorrect results.

If the implementation is hard to explain, it's a bad idea. If the implementation is easy to explain, it may be a good idea.   Many things get more complicated over time: tax laws, romantic relationships, Python programming books. Software is no different. These two aphorisms remind us that if code is so complicated as to be impossible for programmers to understand and debug, it's bad code. But just because it's easy to explain a program's code to someone else doesn't mean it isn't bad code. Unfortunately, figuring out how to make code as simple as possible, and not any simpler, is hard.

Namespaces are one honking great idea—let's do more of those!
Namespaces are separate containers for identifiers to prevent naming conflicts. For example, the open() built-in function and the webbrowser .open() function have the same name but refer to different functions. Importing webbrowser doesn't overwrite the built-in open() function because both open() functions exist in different namespaces: the built-in namespace and the webbrowser module's namespace, respectively. But keep in mind that flat is better than nested: as great as namespaces are, you should make them only to prevent naming conflicts, not to add needless categorization.

As with all opinions about programming, you can argue against those I've listed here, or they might simply be irrelevant to your situation. Arguing over how you should write code or what counts as "pythonic" is rarely as productive as you think it is. (Unless you're writing an entire book full of programming opinions.)

Learning to Love Significant Indentation

The most common concern I hear about Python from programmers coming from other languages is that Python's *significant indentation* (often mistakenly called *significant whitespace*) is weird and unfamiliar. The amount of indentation at the start of a line of code has meaning in Python, because it determines which lines of code are in the same code block.

Grouping blocks of Python code using indentation can seem odd, because other languages begin and end their blocks with braces, { and }. But programmers in non-Python languages usually indent their blocks too, just like Python programmers, to make their code more readable. For example, the Java programming language doesn't have significant indentation. Java programmers don't need to indent blocks of code, but they often do anyway for readability. The following example has a Java function named main() that contains a single call to a println() function:

```java
// Java Example
public static void main(String[] args) {
    System.out.println("Hello, world!");
}
```

This Java code would run just fine if the println() line weren't indented, because the braces, rather than the indentation, are what mark the start and end of blocks in Java. Instead of allowing indentation to be optional, Python forces your code to be consistently readable. But note that Python doesn't have *significant whitespace*, because Python doesn't restrict how you can use nonindentation whitespace (both 2 + 2 and 2+2 are valid Python expressions).

Some programmers argue that the opening brace should be on the same line as the opening statement, while others argue it should be on the following line. Programmers will argue the merits of their preferred style until the end of time. Python neatly sidesteps this issue by not using braces at all, letting Pythonistas get back to more productive work. I've come to wish that all programming languages would adopt Python's approach to grouping blocks of code.

But some people still long for braces and want to add them to a future version of Python—despite how unpythonic they are. Python's __future__ module backports features to earlier Python versions, and you'll find a hidden Easter egg if you try to import a braces feature into Python:

```
>>> from __future__ import braces
SyntaxError: not a chance
```

I wouldn't count on braces being added to Python any time soon.

Commonly Misused Syntax

If Python isn't your first programming language, you might write your Python code with the same strategies you use to write code in other programming languages. Or perhaps you learned an unusual way of writing your Python code because you were unaware that there are more established best practices. This awkward code works, but you could save some time and effort by learning more standard approaches to writing pythonic code. This section explains common missteps programmers make and how you should write the code instead.

Use enumerate() Instead of range()

When looping over a list or other sequence, some programmers use the range() and len() functions to generate the index integers from 0 up to, but not including, the length of the sequence. It's common to use the variable name i (for index) in these for loops. For example, enter the following unpythonic example into the interactive shell:

```
>>> animals = ['cat', 'dog', 'moose']
>>> for i in range(len(animals)):
...     print(i, animals[i])
...
0 cat
1 dog
2 moose
```

The range(len()) convention is straightforward but less than ideal because it can be difficult to read. Instead, pass the list or sequence to the built-in enumerate() function, which will return an integer for the index and the item at that index. For example, you can write the following pythonic code.

```
>>> # Pythonic Example
>>> animals = ['cat', 'dog', 'moose']
>>> for i, animal in enumerate(animals):
...     print(i, animal)
...
0 cat
1 dog
2 moose
```

The code you write will be slightly cleaner using enumerate() instead of range(len()). If you need only the items but not the indexes, you can still directly iterate over the list in a pythonic way:

```
>>> # Pythonic Example
>>> animals = ['cat', 'dog', 'moose']
>>> for animal in animals:
...     print(animal)
...
cat
dog
moose
```

Calling enumerate() and iterating over a sequence directly are preferable to using the old-fashioned range(len()) convention.

Use the with Statement Instead of open() and close()

The open() function will return a file object that contains methods for reading or writing a file. When you're done, the file object's close() method makes the file available to other programs for reading and writing. You can use these functions individually. But doing so is unpythonic. For example, enter the following into the interactive shell to write the text "Hello, world!" to a file named *spam.txt*:

```
>>> # Unpythonic Example
>>> fileObj = open('spam.txt', 'w')
>>> fileObj.write('Hello, world!')
13
>>> fileObj.close()
```

Writing code this way can lead to unclosed files if, say, an error occurs in a try block and the program skips the call to close(). For example:

```
>>> # Unpythonic Example
>>> try:
...     fileObj = open('spam.txt', 'w')
...     eggs = 42 / 0    # A zero divide error happens here.
```

```
...     fileObj.close()  # This line never runs.
... except:
...     print('Some error occurred.')
...
Some error occurred.
```

Upon reaching the zero divide error, the execution moves to the except block, skipping the close() call and leaving the file open. This can lead to file corruption bugs later that are hard to trace back to the try block.

Instead, you can use the with statement to automatically call close() when the execution leaves the with statement's block. The following pythonic example does the same task as the first example in this section:

```
>>> # Pythonic Example
>>> with open('spam.txt', 'w') as fileObj:
...     fileObj.write('Hello, world!')
...
```

Even though there's no explicit call to close(), the with statement will know to call it when the execution leaves the block. .

Use is to Compare with None Instead of ==

The == equality operator compares two object's values, whereas the is identity operator compares two object's identities. Chapter 7 covers value and identity. Two objects can store equivalent values, but being two separate objects means they have separate identities. However, whenever you compare a value to None, you should almost always use the is operator rather than the == operator.

In some cases, the expression spam == None could evaluate to True even when spam merely contains None. This can happen due to overloading the == operator, which Chapter 17 covers in more detail. But spam is None will check whether the value in the spam variable is literally None. Because None is the only value of the NoneType data type, there is only one None object in any Python program. If a variable is set to None, the is None comparison will always evaluate to True. Chapter 17 describes the specifics of overloading the == operator, but the following is an example of this behavior:

```
>>> class SomeClass:
...     def __eq__(self, other):
...         if other is None:
...             return True
...
>>> spam = SomeClass()
>>> spam == None
True
>>> spam is None
False
```

The possibility that a class overloads the == operator this way is rare, but it's become idiomatic Python to always use is None instead of == None just in case.

Finally, you shouldn't use the is operator with the values True and False. You can use the == equality operator to compare a value with True or False, such as spam == True or spam == False. Even more common is to leave out the operator and Boolean value altogether, writing code like if spam: or if not spam: instead of if spam == True: or if spam == False:.

Formatting Strings

Strings appear in almost every computer program, no matter the language. This data type is common, so it's no surprise there are many approaches to manipulating and formatting strings. This section highlights a couple of best practices.

Use Raw Strings If Your String Has Many Backslashes

Escape characters allow you to insert text into string literals that would otherwise be impossible to include. For example, you need the \ in 'Zophie\'s chair' so Python interprets the second quote as part of the string, not the symbol marking the end of the string. Because the backslash has this special escape meaning, if you want to put an actual backslash character in your string, you must enter it as \\.

Raw strings are string literals that have an r prefix, and they don't treat the backslash characters as escape characters. Instead, they just put the backslashes into the string. For example, this string of a Windows file path requires several escaped backslashes, which isn't very pythonic:

```
>>> # Unpythonic Example
>>> print('The file is in C:\\Users\\Al\\Desktop\\Info\\Archive\\Spam')
The file is in C:\Users\Al\Desktop\Info\Archive\Spam
```

This raw string (notice the r prefix) produces the same string value while being more readable:

```
>>> # Pythonic Example
>>> print(r'The file is in C:\Users\Al\Desktop\Info\Archive\Spam')
The file is in C:\Users\Al\Desktop\Info\Archive\Spam
```

Raw strings aren't a different kind of string data type; they're just a convenient way to type string literals that contain several backslash characters. We often use raw strings to type the strings for regular expressions or Windows file paths, which often have several backslash characters in them that would be a pain to escape individually with \\.

Format Strings with F-Strings

String formatting, or *string interpolation*, is the process of creating strings that include other strings and has had a long history in Python. Originally, the + operator could concatenate strings together, but this resulted in code with many quotes and pluses: `'Hello, ' + name + '. Today is ' + day + ' and it is ' + weather + '.'`. The `%s` conversion specifier made the syntax a bit easier: `'Hello, %s. Today is %s and it is %s.' % (name, day, weather)`. Both techniques will insert the strings in the `name`, `day`, and `weather` variables into the string literals to evaluate to a new string value, like this: `'Hello, Al. Today is Sunday and it is sunny.'`.

The `format()` string method adds the *Format Specification Mini-Language* (*https://docs.python.org/3/library/string.html#formatspec*), which involves using `{}` brace pairs in a way similar to the `%s` conversion specifier. However, the method is somewhat convoluted and can produce unreadable code, so I discourage its use.

But as of Python 3.6, *f-strings* (short for *format strings*) offer a more convenient way to create strings that include other strings. Just like how raw strings are prefixed with an r before the first quote, f-strings are prefixed with an f. You can include variable names in between braces in the f-string to insert the strings stored in those variables:

```
>>> name, day, weather = 'Al', 'Sunday', 'sunny'
>>> f'Hello, {name}. Today is {day} and it is {weather}.'
'Hello, Al. Today is Sunday and it is sunny.'
```

The braces can contain entire expressions as well:

```
>>> width, length = 10, 12
>>> f'A {width} by {length} room has an area of {width * length}.'
'A 10 by 12 room has an area of 120.'
```

If you need to use a literal brace inside an f-string, you can escape it with an additional brace:

```
>>> spam = 42
>>> f'This prints the value in spam: {spam}'
'This prints the value in spam: 42'
>>> f'This prints literal curly braces: {{spam}}'
'This prints literal curly braces: {spam}'
```

Because you can put variable names and expressions inline inside the string, your code becomes more readable than using the old ways of string formatting.

All of these different ways to format strings go against the Zen of Python aphorism that there should be one—and preferably only one—obvious way to do something. But f-strings are an improvement to the language (in my opinion), and as the other guideline states, practicality beats purity. If you're writing code for Python 3.6 or later only, use f-strings. If you're writing code that might be run by earlier Python versions, stick to the format() string method or %s conversion specifiers.

Making Shallow Copies of Lists

The *slice* syntax can easily create new strings or lists from existing ones. Enter the following into the interactive shell to see how it works:

```
>>> 'Hello, world!'[7:12] # Create a string from a larger string.
'world'
>>> 'Hello, world!'[:5] # Create a string from a larger string.
'Hello'
>>> ['cat', 'dog', 'rat', 'eel'][2:] # Create a list from a larger list.
['rat', 'eel']
```

The colon (:) separates the starting and ending indexes of the items to put in the new list you're creating. If you omit the starting index before the colon, as in 'Hello, world!'[:5], the starting index defaults to 0. If you omit the ending index after the colon, as in ['cat', 'dog', 'rat', 'eel'][2:], the ending index defaults to the end of the list.

If you omit both indexes, the starting index is 0 (the start of the list) and the ending index is the end of the list. This effectively creates a copy of the list:

```
>>> spam = ['cat', 'dog', 'rat', 'eel']
>>> eggs = spam[:]
>>> eggs
['cat', 'dog', 'rat', 'eel']
>>> id(spam) == id(eggs)
False
```

Notice that the identities of the lists in spam and eggs are different. The eggs = spam[:] line creates a *shallow copy* of the list in spam, whereas eggs = spam would copy only the reference to the list. But the [:] does look a bit odd, and using the copy module's copy() function to produce a shallow copy of the list is more readable:

```
>>> # Pythonic Example
>>> import copy
>>> spam = ['cat', 'dog', 'rat', 'eel']
>>> eggs = copy.copy(spam)
>>> id(spam) == id(eggs)
False
```

You should know about this odd syntax in case you come across Python code that uses it, but I don't recommend writing it in your own code. Keep in mind that both [:] and copy.copy() create shallow copies.

Pythonic Ways to Use Dictionaries

Dictionaries are at the core of many Python programs because of the flexibility that key-value pairs (discussed further in Chapter 7) provide by mapping one piece of data to another. Therefore, it's useful to learn about some of the dictionary idioms Python code commonly uses.

For further information about dictionaries, consult Python programmer Brandon Rhodes's incredible talks about dictionaries and how they work: "The Mighty Dictionary" at PyCon 2010, viewable at *https://invpy.com/mightydictionary*, and "The Dictionary Even Mightier" at PyCon 2017, viewable at *https://invpy.com/dictionaryevenmightier.*

Use get() and setdefault() with Dictionaries

Trying to access a dictionary key that doesn't exist will result in a KeyError error, so programmers will often write unpythonic code to avoid the situation, like this:

```
>>> # Unpythonic Example
>>> numberOfPets = {'dogs': 2}
>>> if 'cats' in numberOfPets: # Check if 'cats' exists as a key.
...     print('I have', numberOfPets['cats'], 'cats.')
... else:
...     print('I have 0 cats.')
...
I have 0 cats.
```

This code checks whether the string 'cats' exists as a key in the numberOfPets dictionary. If it does, a print() call accesses numberOfPets['cats'] as part of a message for the user. If it doesn't, another print() call prints a string without accessing numberOfPets['cats'] so it doesn't raise a KeyError.

This pattern happens so often that dictionaries have a get() method that allows you to specify a default value to return when a key doesn't exist in the dictionary. The following pythonic code is equivalent to the previous example:

```
>>> # Pythonic Example
>>> numberOfPets = {'dogs': 2}
>>> print('I have', numberOfPets.get('cats', 0), 'cats.')
I have 0 cats.
```

The numberOfPets.get('cats', 0) call checks whether the key 'cats' exists in the numberOfPets dictionary. If it does, the method call returns the value for the 'cats' key. If it doesn't, it returns the second argument, 0, instead.

Using the get() method to specify a default value to use for nonexistent keys is shorter and more readable than using if-else statements.

Conversely, you might want to set a default value if a key doesn't exist. For example, if the dictionary in numberOfPets doesn't have a 'cats' key, the instruction numberOfPets['cats'] += 10 would result in a KeyError error. You might want to add code that checks for the key's absence and sets a default value:

```
>>> # Unpythonic Example
>>> numberOfPets = {'dogs': 2}
>>> if 'cats' not in numberOfPets:
...     numberOfPets['cats'] = 0
...
>>> numberOfPets['cats'] += 10
>>> numberOfPets['cats']
10
```

But because this pattern is also common, dictionaries have a more pythonic setdefault() method. The following code is equivalent to the previous example:

```
>>> # Pythonic Example
>>> numberOfPets = {'dogs': 2}
>>> numberOfPets.setdefault('cats', 0) # Does nothing if 'cats' exists.
0
>>> numberOfPets['cats'] += 10
>>> numberOfPets['cats']
10
```

If you're writing if statements that check whether a key exists in a dictionary and sets a default value if the key is absent, use the setdefault() method instead.

Use collections.defaultdict for Default Values

You can use the collections.defaultdict class to eliminate KeyError errors entirely. This class lets you create a default dictionary by importing the collections module and calling collections.defaultdict(), passing it a data type to use for a default value. For example, by passing int to collections.defaultdict(), you can make a dictionary-like object that uses 0 for a default value of nonexistent keys. Enter the following into the interactive shell:

```
>>> import collections
>>> scores = collections.defaultdict(int)
>>> scores
defaultdict(<class 'int'>, {})
>>> scores['Al'] += 1 # No need to set a value for the 'Al' key first.
>>> scores
defaultdict(<class 'int'>, {'Al': 1})
>>> scores['Zophie'] # No need to set a value for the 'Zophie' key first.
```

```
0
>>> scores['Zophie'] += 40
>>> scores
defaultdict(<class 'int'>, {'Al': 1, 'Zophie': 40})
```

Note that you're passing the int() function, not calling it, so you omit the parentheses after int in collections.defaultdict(int). You can also pass list to use an empty list as the default value. Enter the following into the interactive shell:

```
>>> import collections
>>> booksReadBy = collections.defaultdict(list)
>>> booksReadBy['Al'].append('Oryx and Crake')
>>> booksReadBy['Al'].append('American Gods')
>>> len(booksReadBy['Al'])
2
>>> len(booksReadBy['Zophie']) # The default value is an empty list.
0
```

If you need a default value for every possible key, it's much easier to use collections.defaultdict() than use a regular dictionary and constantly call the setdefault() method.

Use Dictionaries Instead of a switch Statement

Languages such as Java have a switch statement, which is a kind of if-elif-else statement that runs code based on which one of many values a specific variable contains. Python doesn't have a switch statement, so Python programmers sometimes write code like the following example, which runs a different assignment statement based on which one of many values the season variable contains:

```
# All of the following if and elif conditions have "season ==":
if season == 'Winter':
    holiday = 'New Year\'s Day'
elif season == 'Spring':
    holiday = 'May Day'
elif season == 'Summer':
    holiday = 'Juneteenth'
elif season == 'Fall':
    holiday = 'Halloween'
else:
    holiday = 'Personal day off'
```

This code isn't necessarily unpythonic, but it's a bit verbose. By default, Java switch statements have "fall-through" that requires each block to end with a break statement. Otherwise, the execution continues on to the next block. Forgetting to add this break statement is a common source of bugs. But all the if-elif statements in our Python example can be repetitive.

Some Python programmers prefer to set up a dictionary value instead of using if-elif statements. The following concise and pythonic code is equivalent to the previous example:

```
holiday = {'Winter': 'New Year\'s Day',
           'Spring': 'May Day',
           'Summer': 'Juneteenth',
           'Fall':   'Halloween'}.get(season, 'Personal day off')
```

This code is just a single assignment statement. The value stored in holiday is the return value of the get() method call, which returns the value for the key that season is set to. If the season key doesn't exist, get() returns 'Personal day off'. Using a dictionary will result in more concise code, but it can also make your code harder to read. It's up to you whether or not to use this convention.

Conditional Expressions: Python's "Ugly" Ternary Operator

Ternary operators (officially called *conditional expressions*, or sometimes *ternary selection expressions*, in Python) evaluate an expression to one of two values based on a condition. Normally, you would do this with a pythonic if-else statement:

```
>>> # Pythonic Example
>>> condition = True
>>> if condition:
...     message = 'Access granted'
... else:
...     message = 'Access denied'
...
>>> message
'Access granted'
```

Ternary simply means an operator with three inputs, but in programming it's synonymous with *conditional expression*. Conditional expressions also offer a more concise one-liner for code that fits this pattern. In Python, they're implemented with an odd arrangement of the if and else keywords:

```
>>> valueIfTrue = 'Access granted'
>>> valueIfFalse = 'Access denied'
>>> condition = True
❶ >>> message = valueIfTrue if condition else valueIfFalse
>>> message
'Access granted'
❷ >>> print(valueIfTrue if condition else valueIfFalse)
'Access granted'
>>> condition = False
>>> message = valueIfTrue if condition else valueIfFalse
>>> message
'Access denied'
```

The expression `valueIfTrue if condition else valueIfFalse` ❶ evaluates to `valueIfTrue` if the `condition` variable is `True`. When the `condition` variable is `False`, the expression evaluates to `valueIfFalse`. Guido van Rossum jokingly described his syntax design as "intentionally ugly." Most languages with a ternary operator list the condition first, followed by the true value and then the false value. You can use a conditional expression anywhere you can use an expression or value, including as the argument to a function call ❷.

Why would Python introduce this syntax in Python 2.5 even though it breaks the first guideline that beautiful is better than ugly? Unfortunately, despite being somewhat unreadable, many programmers use ternary operators and wanted Python to support this syntax. It's possible to abuse Boolean operator short-circuiting to create a *sort of* ternary operator. The expression `condition and valueIfTrue or valueIfFalse` will evaluate to `valueIfTrue` if `condition` is `True`, and `valueIfFalse` if `condition` is `False` (except in one important case). Enter the following into the interactive shell:

```
>>> # Unpythonic Example
>>> valueIfTrue = 'Access granted'
>>> valueIfFalse = 'Access denied'
>>> condition = True
>>> condition and valueIfTrue or valueIfFalse
'Access granted'
```

This `condition and valueIfTrue or valueIfFalse` style of pseudo-ternary operator has a subtle bug: if `valueIfTrue` is a falsey value (such as 0, `False`, `None`, or the blank string), the expression unexpectedly evaluates to `valueIfFalse` if `condition` is `True`.

But programmers continued to use this fake ternary operator anyway, and "Why doesn't Python have a ternary operator?" became a perennial question to the Python core developers. Conditional expressions were created so programmers would stop asking for a ternary operator and wouldn't use the bug-prone pseudo-ternary operator. But conditional expressions are also ugly enough to discourage programmers from using them. Although beautiful may be better than ugly, Python's "ugly" ternary operator is an example of a case when practicality beats purity.

Conditional expressions aren't exactly pythonic, but they're not unpythonic, either. If you do use them, avoid nesting conditional expressions inside other conditional expressions:

```
>>> # Unpythonic Example
>>> age = 30
>>> ageRange = 'child' if age < 13 else 'teenager' if age >= 13 and age < 18
else 'adult'
>>> ageRange
'adult'
```

Nested conditional expressions are a good example of how a dense one-liner can be technically correct but frustrating to make sense of when reading.

Working with Variable Values

You'll often need to check and modify the values that variables store. Python has several ways of doing this. Let's look at a couple of examples.

Chaining Assignment and Comparison Operators

When you have to check whether a number is within a certain range, you might use the Boolean and operator like this:

```
# Unpythonic Example
if 42 < spam and spam < 99:
```

But Python lets you chain comparison operators so you don't need to use the and operator. The following code is equivalent to the previous example:

```
# Pythonic Example
if 42 < spam < 99:
```

The same applies to chaining the = assignment operator. You can set multiple variables to the same value in a single line of code:

```
>>> # Pythonic Example
>>> spam = eggs = bacon = 'string'
>>> print(spam, eggs, bacon)
string string string
```

To check whether all three of these variables are the same, you can use the and operator, or more simply, chain the == comparison operator for equality.

```
>>> # Pythonic Example
>>> spam = eggs = bacon = 'string'
>>> spam == eggs == bacon == 'string'
True
```

Chaining operators is a small but useful shortcut in Python. However, if you use them incorrectly, they can cause problems. Chapter 8 demonstrates some instances where using them can introduce unexpected bugs in your code.

Checking Whether a Variable Is One of Many Values

You might sometimes encounter the inverse of the situation described in the preceding section: checking whether a single variable is one of multiple possible values. You could do this using the or operator, such as in the expression spam == 'cat' or spam == 'dog' or spam == 'moose'. All of those redundant "spam ==" parts make this expression a bit unwieldy.

Instead, you can put the multiple values into a tuple and check for whether a variable's value exists in that tuple using the in operator, as in the following example:

```
>>> # Pythonic Example
>>> spam = 'cat'
>>> spam in ('cat', 'dog', 'moose')
True
```

Not only is this idiom easier to understand, it's also slightly faster, according to timeit.

Summary

All programming languages have their own idioms and best practices. This chapter focuses on the particular ways that Python programmers have come to write "pythonic" code to make the best use of Python's syntax.

At the core of pythonic code are the 20 aphorisms from the Zen of Python, which are rough guidelines for writing Python. These aphorisms are opinions and not strictly necessary for writing Python code, but they are good to keep in mind.

Python's significant indentation (not to be confused with significant whitespace) provokes the most protest from new Python programmers. Although almost all programming languages commonly use indentation to make code readable, Python requires it in place of the more typical braces that other languages use.

Although many Python programmers use the range(len()) convention for for loops, the enumerate() function offers a cleaner approach to getting the index and value while iterating over a sequence. Similarly, the with statement is a cleaner and less bug-prone way to handle files compared to calling open() and close() manually. The with statement ensures that close() gets called whenever the execution moves outside the with statement's block.

Python has had several ways to interpolate strings. The original way was to use the %s conversion specifier to mark where strings should be included in the original string. The modern way as of Python 3.6 is to use f-strings. F-strings prefix the string literal with the letter f and use braces to mark where you can place strings (or entire expressions) inside the string.

The [:] syntax for making shallow copies of lists is a bit odd-looking and not necessarily pythonic, but it's become a common way to quickly create a shallow list.

Dictionaries have a get() and setdefault() method for dealing with nonexistent keys. Alternatively, a collections.defaultdict dictionary will use a default value for nonexistent keys. Also, although there is no switch statement in Python, using a dictionary is a terse way to implement its equivalent without using several if-elif-else statements, and you can use ternary operators when evaluating between two values.

A chain of == operators can check whether multiple variables are equal to each other, whereas the in operator can check whether a variable is one of many possible values.

This chapter covered several Python language idioms, providing you with hints for how to write more pythonic code. In the next chapter, you'll learn about some of the Python gotchas and pitfalls that beginners fall into.

7

PROGRAMMING JARGON

In the XKCD comic "Up Goer Five" (*https://xkcd.com/1133/*), the webcomic's artist Randall Munroe created a technical schematic for the Saturn V rocket using only the 1,000 most common English words. The comic breaks down all the technical jargon into sentences a young child could understand. But it also highlights why we can't explain everything using simple terms: The explanation "Thing to help people escape really fast if there's a problem and everything is on fire so they decide not to go to space" might be easier to understand for a lay audience than "Launch Escape System." But it's too verbose for NASA engineers to say in their day-to-day work. Even then, they'd probably rather use the acronym LES.

Although computer jargon can be confusing and intimidating for new programmers, it's a necessary shorthand. Several terms in Python and software development have subtle differences in meaning, and even experienced developers sometimes carelessly use them interchangeably. The technical definitions for these terms can vary between programming languages, but this chapter covers the terms as they relate to Python. You'll get a broad, albeit not deep, understanding of the programming language concepts behind them.

This chapter assumes you aren't yet familiar with classes and object-oriented programming (OOP). I've limited the explanations for classes and other OOP jargon here, but the jargon is explained in more detail in Chapters 15 to 17.

Definitions

As the number of programmers in a room approaches two, the likelihood of an argument about semantics approaches 100 percent. Language is fluid and humans are the masters of words rather than the other way around. Some developers might use terms slightly differently, but becoming familiar with these terms is still useful. This chapter explores these terms and how they compare with each other. If you need a glossary of terms in alphabetical order, you can rely on the official Python glossary at *https://docs.python.org/3/glossary.html* to provide canonical definitions.

No doubt, some programmers will read the definitions in this chapter and bring up special cases or exceptions that can be endlessly nitpicked. Rather than being a definitive guide, this chapter is intended to give you accessible definitions, even if they're not comprehensive. As with everything in programming, there's always more to learn.

Python the Language and Python the Interpreter

The word *python* can have multiple meanings. The Python programming language gets its name from the British comedy group Monty Python, rather than the snake (although Python tutorials and documentation use both Monty Python and snake references). Similarly, *Python* can have two meanings in regard to computer programming.

When we say, "Python runs a program" or "Python will raise an exception," we're talking about the *Python interpreter*—the actual software that reads the text of a *.py* file and carries out its instructions. When we say, "the Python interpreter," we're almost always talking about *CPython,* the Python interpreter maintained by the Python Software Foundation, available at *https://www.python.org.* CPython is an *implementation* of the Python language—that is, software created to follow a specification—but there are others. Although CPython is written in the C programming language, *Jython* is written in Java for running Python scripts that are interoperable with Java programs. *PyPy,* a *just-in-time compiler* for Python that compiles as programs execute, is written in Python.

All of these implementations run source code written in the *Python programming language*, which is what we mean when we say, "This is a Python program" or "I'm learning Python." Ideally, any Python interpreter can run any source code written in the Python language; however, in the real world there'll be some slight incompatibilities and differences between interpreters. CPython is called the Python language's *reference implementation* because if there's a difference between how CPython and another interpreter interpret Python code, CPython's behavior is considered canonical and correct.

Garbage Collection

In many early programming languages, a programmer had to instruct the program to allocate and then deallocate, or free, memory for data structures as needed. Manual memory allocation was the source of numerous bugs, such as *memory leaks* (where programmers forgot to free memory) or *double-free bugs* (where programmers freed the same memory twice, leading to data corruption).

To avoid these bugs, Python has *garbage collection*, a form of automatic memory management that tracks when to allocate and free memory so the programmer doesn't have to. You can think of garbage collection as memory recycling, because it makes memory available for new data. For example, enter the following into the interactive shell:

```
>>> def someFunction():
...     print('someFunction() called.')
...     spam = ['cat', 'dog', 'moose']
...
>>> someFunction()
someFunction() called.
```

When someFunction() is called, Python allocates memory for the list ['cat', 'dog', 'moose']. The programmer doesn't need to figure out how many bytes of memory to request because Python manages this automatically. Python's garbage collector will free the local variables when the function call returns to make that memory available for other data. Garbage collection makes programming much easier and less bug-prone.

Literals

A *literal* is text in the source code for a fixed, typed-out value. In the following code example

```
>>> age = 42 + len('Zophie')
```

the 42 and 'Zophie' text are integer and string literals. Think of a literal as a value that literally appears in source code text. Only the built-in data types can have literal values in Python source code, so the variable age isn't a literal value. Table 7-1 lists some example Python literals.

Table 7-1: Examples of Literals in Python

Literal	Data type
42	Integer
3.14	Float
1.4886191506362924e+36	Float
"""Howdy!"""	String
r'Green\Blue'	String
[]	List
{'name': 'Zophie'}	Dictionary
b'\x41'	Bytes
True	Boolean
None	NoneType

Nitpickers will argue that some of my choices aren't literals based on the official Python language documentation. Technically, -5 isn't a literal in Python because the language defines the negative symbol (-) as an operator that operates on the 5 literal. In addition, True, False, and None are considered Python keywords rather than literals, whereas [] and {} are called *displays* or *atoms* depending on what part of the official documentation you're looking at. Regardless, literal is a common term that software professionals will use for all of these examples.

Keywords

Every programming language has its own *keywords*. The Python keywords are a set of names reserved for use as part of the language and cannot be used as variable names (that is, as identifiers). For example, you cannot have a variable named while because while is a keyword reserved for use in while loops. The following are the Python keywords as of Python 3.9.

and	continue	finally	is	raise
as	def	for	lambda	return
assert	del	from	None	True
async	elif	global	nonlocal	try
await	else	if	not	while
break	except	import	or	with
class	False	in	pass	yield

Note that the Python keywords are always in English and aren't available in alternative languages. For example, the following function has identifiers written in Spanish, but the def and return keywords remain in English.

```
def agregarDosNúmeros(primerNúmero, segundoNúmero):
    return primerNúmero + segundoNúmero
```

Unfortunately for the 6.5 billion people who don't speak it, English dominates the programming field.

Objects, Values, Instances, and Identities

An *object* is a representation of a piece of data, such as a number, some text, or a more complicated data structure, such as a list or dictionary. All objects can be stored in variables, passed as arguments to function calls, and returned from function calls.

All objects have a value, identity, and data type. The *value* is the data the object represents, such as the integer 42 or the string 'hello'. Although somewhat confusing, some programmers use the term *value* as a synonym for *object*, especially for simple data types like integers or strings. For example, a variable that contains 42 is a variable that contains an integer value, but we can also say it's a variable that contains an integer object with a value of 42.

An object is created with an *identity* that is a unique integer you can view by calling the id() function. For example, enter the following code into the interactive shell:

```
>>> spam = ['cat', 'dog', 'moose']
>>> id(spam)
33805656
```

The variable spam stores an object of the list data type. Its value is ['cat', 'dog', 'moose']. Its identity is 33805656, although the integer ID varies each time a program runs so you'll likely get a different ID on your computer. Once created, an object's identity won't change for as long as the program runs. Although the data type and the object's identity will never change, an object's value can change, as we'll see in this example:

```
>>> spam.append('snake')
>>> spam
['cat', 'dog', 'moose', 'snake']
>>> id(spam)
33805656
```

Now the list also contains 'snake'. But as you can see from the id(spam) call, its identity hasn't changed and it's still the same list. But let's see what happens when you enter this code:

```
>>> spam = [1, 2, 3]
>>> id(spam)
33838544
```

The value in spam has been overwritten by a new list object with a new identity: 33838544 instead of 33805656. An *identifier* like spam isn't the same as an *identity* because multiple identifiers can refer to the same object, as is the case in this example of two variables that are assigned to the same dictionary:

```
>>> spam = {'name': 'Zophie'}
>>> id(spam)
33861824
>>> eggs = spam
>>> id(eggs)
33861824
```

The identities of the spam and eggs identifiers are both 33861824 because they refer to the same dictionary object. Now change the value of spam in the interactive shell:

```
    >>> spam = {'name': 'Zophie'}
    >>> eggs = spam
 ❶ >>> spam['name'] = 'Al'
    >>> spam
    {'name': 'Al'}
    >>> eggs
 ❷ {'name': 'Al'}
```

You'll see that changes to spam ❶ mysteriously also appear in eggs ❷. The reason is that they both refer to the same object.

VARIABLE METAPHORS: BOX VS. LABEL

Many introductory books use boxes as a metaphor for variables, which is an oversimplification. It's easy to think of variables as a box that a value is stored in, as in Figure 7-1, but this metaphor falls apart when it comes to references. The previous spam and eggs variables don't store separate dictionaries; rather, they store *references* to the same dictionary in the computer's memory.

Figure 7-1: Many books say you can think of a variable as a box that contains a value.

In Python, all variables are technically references, not containers of values, regardless of their data type. The box metaphor is simple but also flawed. Instead of thinking of variables as boxes, you can think of variables as labels for objects in memory. Figure 7-2 shows labels on the previous spam and eggs examples.

Figure 7-2: Variables can also be thought of as labels on values.

Because multiple variables can refer to the same object, that object can be "stored" in multiple variables. Multiple boxes can't store the same object, so it might be easier for you to use the label metaphor instead. Ned Batchelder's PyCon 2015 talk, "Facts and Myths about Python Names and Values" has more information on this topic at *https://youtu.be/_AEJHKGk9ns*.

Without understanding that the = assignment operator always copies the reference, not the object, you might introduce bugs by thinking that you're making a duplicate copy of an object when really you're copying the reference to the original object. Fortunately, this isn't an issue for immutable values like integers, strings, and tuples for reasons that I'll explain in "Mutable and Immutable" on page 114.

You can use the is operator to compare whether two objects have the same identity. In contrast, the == operator checks only whether object values are the same. You can consider x is y to be shorthand for id(x) == id(y). Enter the following into the interactive shell to see the difference:

```
>>> spam = {'name': 'Zophie'}
❶ >>> eggs = spam
>>> spam is eggs
True
>>> spam == eggs
True
❷ >>> bacon = {'name': 'Zophie'}
>>> spam == bacon
True
>>> spam is bacon
False
```

The variables spam and eggs refer to the same dictionary object ❶, so their identities and values are the same. But bacon refers to a separate dictionary object ❷, even though it contains data identical to spam and eggs. The identical data means bacon has the same value as spam and eggs, but they're two different objects with two different identities.

Items

In Python, an object that is inside a container object, like a list or dictionary, is also called an *item* or an *element*. For example, the strings in the list ['dog', 'cat', 'moose'] are objects but are also called items.

Mutable and Immutable

As noted earlier, all objects in Python have a value, data type, and identity, and of these only the value can change. If you can change the object's value, it's a *mutable* object. If you can't change its value, it's an *immutable* object. Table 7-2 lists some mutable and immutable data types in Python.

Table 7-2: Some of Python's Mutable and Immutable Data Types

Mutable data types	Immutable data types
List	Integer
Dictionaries	Floating-point number
Sets	Boolean
Bytearray	String
Array	Frozen set
	Bytes
	Tuple

When you overwrite a variable, it might look like you're changing the object's value, as in this interactive shell example:

```
>>> spam = 'hello'
>>> spam
'hello'
>>> spam = 'goodbye'
>>> spam
'goodbye'
```

But in this code, you haven't changed the 'hello' object's value from 'hello' to 'goodbye'. They're two separate objects. You've only switched spam from referring to the 'hello' object to the 'goodbye' object. You can check whether this is true by using the id() function to show the two objects' identities:

```
>>> spam = 'hello'
>>> id(spam)
```

```
40718944
>>> spam = 'goodbye'
>>> id(spam)
40719224
```

These two string objects have different identities (40718944 and 40719224) because they're different objects. But variables that refer to mutable objects can have their values modified *in-place*. For example, enter the following into the interactive shell:

```
>>> spam = ['cat', 'dog']
>>> id(spam)
33805576
❶ >>> spam.append('moose')
❷ >>> spam[0] = 'snake'
>>> spam
['snake', 'dog', 'moose']
>>> id(spam)
33805576
```

The append() method ❶ and item assignment by indexing ❷ both modify the value of the list in-place. Even though the list's *value* has changed, its *identity* remains the same (33805576). But when you concatenate a list using the + operator, you create a new object (with a new identity) that overwrites the old list:

```
>>> spam = spam + ['rat']
>>> spam
['snake', 'dog', 'moose', 'rat']
>>> id(spam)
33840064
```

List concatenation creates a new list with a new identity. When this happens, the old list will eventually be freed from memory by the garbage collector. You'll have to consult the Python documentation to see which methods and operations modify objects in-place and which overwrite objects. A good rule to keep in mind is that if you see a literal in the source code, such as ['rat'] in the previous example, Python will most likely create a new object. A method that is called on the object, such as append(), often modifies the object in-place.

Assignment is simpler for objects of immutable data types like integers, strings, or tuples. For example, enter the following into the interactive shell:

```
>>> bacon = 'Goodbye'
>>> id(bacon)
33827584
❶ >>> bacon = 'Hello'
>>> id(bacon)
33863820
❷ >>> bacon = bacon + ', world!'
>>> bacon
'Hello, world!'
```

```
>>> id(bacon)
33870056
```
❸ `>>> bacon[0] = 'J'`
```
Traceback (most recent call last):
  File "<stdin>", line 1, in <module>
TypeError: 'str' object does not support item assignment
```

Strings are immutable, so you cannot change their value. Although it looks like the string's value in bacon is being changed from 'Goodbye' to 'Hello' ❶, it's actually being overwritten by a new string object with a new identity. Similarly, an expression using string concatenation creates a new string object ❷ with a new identity. Attempting to modify the string in-place with item assignment isn't allowed in Python ❸.

A tuple's value is defined as the objects it contains and the order of those objects. *Tuples* are immutable sequence objects that enclose values in parentheses. This means that items in a tuple can't be overwritten:

```
>>> eggs = ('cat', 'dog', [2, 4, 6])
>>> id(eggs)
39560896
>>> id(eggs[2])
40654152
>>> eggs[2] = eggs[2] + [8, 10]
Traceback (most recent call last):
  File "<stdin>", line 1, in <module>
TypeError: 'tuple' object does not support item assignment
```

But a mutable list inside an immutable tuple can still be modified in-place:

```
>>> eggs[2].append(8)
>>> eggs[2].append(10)
>>> eggs
('cat', 'dog', [2, 4, 6, 8, 10])
>>> id(eggs)
39560896
>>> id(eggs[2])
40654152
```

Although this is an obscure special case, it's important to keep in mind. The tuple still refers to the same objects, as depicted in Figure 7-3. But if a tuple contains a mutable object and that object changes its value—that is, if the object mutates—the value of the tuple also changes.

I, and almost every Pythonista, call tuples immutable. But whether some tuples can be called mutable depends on your definition. I explore this topic more in my PyCascades 2019 talk, "The Amazing Mutable, Immutable Tuple" at *https://inupy.com/amazingtuple/*. You can also read Luciano Ramalho's explanation in Chapter 2 of *Fluent Python*. (O'Reilly Media, 2015)

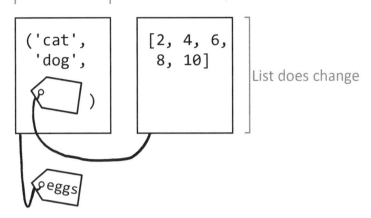

Tuple doesn't change

List does change

Figure 7-3: Although the set of objects in a tuple is immutable, the objects can be mutable.

Indexes, Keys, and Hashes

Python lists and dictionaries are values that can contain multiple other values. To access these values, you use an *index operator*, which is composed of a pair of square brackets ([ ]) and an integer called an *index* to specify which value you want to access. Enter the following into the interactive shell to see how indexing works with lists:

```
>>> spam = ['cat', 'dog', 'moose']
>>> spam[0]
'cat'
>>> spam[-2]
'dog'
```

In this example, 0 is an index. The first index is 0, not 1, because Python (as most languages do) uses *zero-based indexing*. Languages that use one-based indexing are rare: Lua and R are the most predominant. Python also supports negative indexes, where -1 refers to the last item in a list, -2 refers to the second-to-last item, and so on. You can think of a negative index spam[-n] as being the same as spam[len(spam) - n].

NOTE   *Computer scientist and singer-songwriter Stan Kelly-Bootle once joked, "Should array indices start at 0 or 1? My compromise of 0.5 was rejected without, I thought, proper consideration."*

You can also use the index operator on a list literal, although all those square brackets can look confusing and unnecessary in real-world code:

```
>>> ['cat', 'dog', 'moose'][2]
'moose'
```

Indexing can also be used for values other than lists, such as on a string to obtain individual characters:

```
>>> 'Hello, world'[0]
'H'
```

Python dictionaries are organized into *key-value pairs*:

```
>>> spam = {'name': 'Zophie'}
>>> spam['name']
'Zophie'
```

Although list indexes are limited to integers, a Python dictionary's index operator is a *key* and can be any hashable object. A *hash* is an integer that acts as a sort of fingerprint for a value. An object's hash never changes for the lifetime of the object, and objects with the same value must have the same hash. The string 'name' in this instance is the key for the value 'Zophie'. The hash() function will return an object's hash if the object is *hashable*. Immutable objects, such as strings, integers, floats, and tuples, can be hashable. Lists (as well as other mutable objects) aren't hashable. Enter the following into the interactive shell:

```
>>> hash('hello')
-1734230105925061914
>>> hash(42)
42
>>> hash(3.14)
322818021289917443
>>> hash((1, 2, 3))
2528502973977326415
>>> hash([1, 2, 3])
Traceback (most recent call last):
  File "<stdin>", line 1, in <module>
TypeError: unhashable type: 'list'
```

Although the details are beyond the scope of this book, the key's hash is used to find items stored in a dictionary and set data structures. That's why you can't use a mutable list for a dictionary's keys:

```
>>> d = {}
>>> d[[1, 2, 3]] = 'some value'
Traceback (most recent call last):
  File "<stdin>", line 1, in <module>
TypeError: unhashable type: 'list'
```

A hash is different from an identity. Two different objects with the same value will have different identities but the same hash. For example, enter the following into the interactive shell:

```
>>> a = ('cat', 'dog', 'moose')
>>> b = ('cat', 'dog', 'moose')
>>> id(a), id(b)
```

```
   (37111992, 37112136)
❶ >>> id(a) == id(b)
   False
   >>> hash(a), hash(b)
   (-3478972040190420094, -3478972040190420094)
❷ >>> hash(a) == hash(b)
   True
```

The tuples referred to by a and b have different identities ❶, but their identical values mean they'll have identical hashes ❷. Note that a tuple is hashable if it contains only hashable items. Because you can use only hashable items as keys in a dictionary, you can't use a tuple that contains an unhashable list as a key. Enter the following into the interactive shell:

```
   >>> tuple1 = ('cat', 'dog')
   >>> tuple2 = ('cat', ['apple', 'orange'])
   >>> spam = {}
❶ >>> spam[tuple1] = 'a value'
❷ >>> spam[tuple2] = 'another value'
   Traceback (most recent call last):
     File "<stdin>", line 1, in <module>
   TypeError: unhashable type: 'list'
```

Notice that tuple1 is hashable ❶, but tuple2 contains an unhashable list ❷ and so is also unhashable.

Containers, Sequences, Mapping, and Set Types

The words *container, sequence,* and *mapping* have meanings in Python that don't necessarily apply to other programming languages. In Python, a *container* is an object of any data type that can contain multiple other objects. Lists and dictionaries are common container types used in Python.

A *sequence* is an object of any container data type with ordered values accessible through integer indexes. Strings, tuples, lists, and bytes objects are sequence data types. Objects of these types can access values using integer indexes in the index operator (the [and] brackets) and can also be passed to the len() function. By "ordered," we mean that there is a first value, second value, and so on in the sequence. For example, the following two list values aren't considered equal because their values are ordered differently:

```
   >>> [1, 2, 3] == [3, 2, 1]
   False
```

A *mapping* is an object of any container data type that uses keys instead of an index. A mapping can be ordered or unordered. Dictionaries in Python 3.4 and earlier are unordered because there is no first or last key-value pair in a dictionary:

```
   >>> spam = {'a': 1, 'b': 2, 'c': 3, 'd': 4}  # This is run from CPython 3.5.
   >>> list(spam.keys())
   ['a', 'c', 'd', 'b']
   >>> spam['e'] = 5
```

```
>>> list(spam.keys())
['e', 'a', 'c', 'd', 'b']
```

You have no guarantee of getting items in a consistent order from dictionaries in early versions of Python. As a result of dictionaries' unordered nature, two dictionary literals written with different orders for their key-value pairs are still considered equal:

```
>>> {'a': 1, 'b': 2, 'c': 3} == {'c': 3, 'a': 1, 'b': 2}
True
```

But starting in CPython 3.6, dictionaries do retain the insertion order of their key-value pairs:

```
>>> spam = {'a': 1, 'b': 2, 'c': 3, 'd': 4}  # This is run from CPython 3.6.
>>> list(spam)
['a', 'b', 'c', 'd']
>>> spam['e'] = 5
>>> list(spam)
['a', 'b', 'c', 'd', 'e']
```

This is a feature in the CPython 3.6 interpreter but not in other interpreters for Python 3.6. All Python 3.7 interpreters support ordered dictionaries, which became standard in the Python language in 3.7. But just because a dictionary is ordered doesn't mean that its items are accessible through integer indexes: spam[0] won't evaluate to the first item in an ordered dictionary (unless by coincidence there is a key 0 for the first item). Ordered dictionaries are also considered the same if they contain the same key-value pairs, even if the key-value pairs are in a different order in each dictionary.

The collections module contains many other mapping types, including OrderedDict, ChainMap, Counter, and UserDict, which are described in the online documentation at *https://docs.python.org/3/library/collections.html*.

Dunder Methods and Magic Methods

Dunder methods, also called *magic methods*, are special methods in Python whose names begin and end with two underscores. These methods are used for operator overloading. *Dunder* is short for double underscore. The most familiar dunder method is __init__() (pronounced "dunder init dunder," or simply "init"), which initializes objects. Python has a few dozen dunder methods, and Chapter 17 explains them in detail.

Modules and Packages

A *module* is a Python program that other Python programs can import so they can use the module's code. The modules that come with Python are collectively called the *Python Standard Library*, but you can create your own modules as well. If you save a Python program as, for example, *spam.py*, other programs can run `import spam` to access the *spam.py* program's functions, classes, and top-level variables.

A *package* is a collection of modules that you form by placing a file named *__init__.py* inside a folder. You use the folder's name as the name of the package. Packages can contain multiple modules (that is, *.py* files) or other packages (other folders containing *__init__.py* files).

For more explanation and detail about modules and packages, check out the official Python documentation at *https://docs.python.org/3/tutorial/modules.html*.

Callables and First-Class Objects

Functions and methods aren't the only things that you can call in Python. Any object that implements the *callable operator*—the two parentheses ()— is a *callable* object. For example, if you have a `def hello():` statement, you can think of the code as a variable named `hello` that contains a function object. Using the callable operator on this variable calls the function in the variable: `hello()`.

Classes are an OOP concept, and a class is an example of a callable object that isn't a function or method. For example, the `date` class in the `datetime` module is called using the callable operator, as in the code `datetime.date(2020, 1, 1)`. When the class object is called, the code inside the class's __init__() method is run. Chapter 15 has more details about classes.

Functions are *first-class objects* in Python, meaning you can store them in variables, pass them as arguments in function calls, return them from function calls, and do anything else you can do with an object. Think of a `def` statement as assigning a function object to a variable. For example, you could create a `spam()` function that you can then call:

```
>>> def spam():
...     print('Spam! Spam! Spam!')
...
>>> spam()
Spam! Spam! Spam!
```

You can also assign the `spam()` function object to other variables. When you call the variable you've assigned the function object to, Python executes the function:

```
>>> eggs = spam
>>> eggs()
Spam! Spam! Spam!
```

These are called *aliases*, which are different names for existing functions. They're often used if you need to rename a function. But a large amount of existing code uses the old name, and it would be too much work to change it.

The most common use of first-class functions is so you can pass functions to other functions. For example, we can define a `callTwice()` function, which can be passed a function that needs to be called twice:

```
>>> def callTwice(func):
...     func()
...     func()
...
>>> callTwice(spam)
Spam! Spam! Spam!
Spam! Spam! Spam!
```

You could just write `spam()` twice in your source code. But you can pass the `callTwice()` function to any function at runtime rather than having to type the function call twice into the source code beforehand.

Commonly Confused Terms

Technical jargon is confusing enough, especially for terms that have related but distinct definitions. To make matters worse, languages, operating systems, and fields in computing might use different terms to mean the same thing or the same terms to mean different things. To communicate clearly with other programmers, you'll need to learn the difference between the following terms.

Statements vs. Expressions

Expressions are the instructions made up of operators and values that evaluate to a single value. A value can be a variable (which contains a value) or a function call (which returns a value). So, `2 + 2` is an expression that evaluates down to the single value of `4`. But `len(myName) > 4` and `myName.isupper()` or `myName == 'Zophie'` are expressions as well. A value by itself is also an expression that evaluates to itself.

Statements are, effectively, all other instructions in Python. These include `if` statements, `for` statements, `def` statements, `return` statements, and so on. Statements do *not* evaluate to a value. Some statements can include expressions, such as an assignment statement like `spam = 2 + 2` or an `if` statement like `if myName == 'Zophie':`.

Although Python 3 uses a `print()` function, Python 2 instead has a `print` statement. The difference might seem like just the introduction of parentheses, but it's important to note that the Python 3 `print()` function has a return value (which is always `None`), can be passed as an argument to other functions, and can be assigned to a variable. None of these actions are possible with statements. However, you can still use the parentheses in Python 2, as in the following interactive shell example:

```
>>> print 'Hello, world!' # run in Python 2
Hello, world!
❶ >>> print('Hello, world!') # run in Python 2
Hello, world!
```

Although this looks like a function call ❶, it's actually a `print` statement with a string value wrapped in parentheses, the same way assigning spam = (2 + 2) is equivalent to spam = 2 + 2. In Python 2 and 3, you can pass multiple values to the `print` statement or `print()` function, respectively. In Python 3, this would look like the following:

```
>>> print('Hello', 'world') # run in Python 3
Hello world
```

But using this same code in Python 2 would be interpreted as passing a tuple of two string values in a `print` statement, producing this output:

```
>>> print('Hello', 'world') # run in Python 2
('Hello', 'world')
```

A statement and an expression composed of a function call have subtle but real differences.

Block vs. Clause vs. Body

The terms *block, clause,* and *body* are often used interchangeably to refer to a group of Python instructions. A *block* begins with indentation and ends when that indentation returns to the previous indent level. For example, the code that follows an `if` or `for` statement is called the statement's block. A new block is required following statements that end with a colon, such as `if`, `else`, `for`, `while`, `def`, `class`, and so on.

But Python does allow one-line blocks. This is valid, although not recommended, Python syntax:

```
if name == 'Zophie': print('Hello, kitty!')
```

By using the semicolon, you can also have multiple instructions in the `if` statement's block:

```
if name == 'Zophie': print('Hello, kitty!'); print('Do you want a treat?')
```

But you can't have one-liners with other statements that require new blocks. The following isn't valid Python code:

```
if name == 'Zophie': if age < 2: print('Hello, kitten!')
```

This is invalid because if an `else` statement is on the next line, it would be ambiguous as to which `if` statement the `else` statement would refer to.

The official Python documentation prefers the term *clause* rather than block (*https://docs.python.org/3/reference/compound_stmts.html*). The following code is a clause:

```
if name == 'Zophie':
    print('Hello, kitty!')
    print('Do you want a treat?')
```

The if statement is the *clause header*, and the two print() calls nested in the if are the *clause suite* or *body*. The official Python documentation uses *block* to refer to a piece of Python code that executes as a unit, such as a module, a function, or a class definition (*https://docs.python.org/3/reference/executionmodel.html*).

Variable vs. Attribute

Variables are simply names that refer to objects. *Attributes* are, to quote the official documentation, "any name following a dot" (*https://docs.python.org/3/tutorial/classes.html#python-scopes-and-namespaces*). Attributes are associated with objects (the name before the dot/period). For example, enter the following into the interactive shell:

```
>>> import datetime
>>> spam = datetime.datetime.now()
>>> spam.year
2018
>>> spam.month
1
```

In this code example, spam is a variable that contains a datetime object (returned from datetime.datetime.now()), and year and month are attributes of that object. Even in the case of, say, sys.exit(), the exit() function is considered an attribute of the sys module object.

Other languages call attributes *fields*, *properties*, or *member variables*.

Function vs. Method

A *function* is a collection of code that runs when called. A *method* is a function (or a *callable*, described in the next section) that is associated with a class, just as an attribute is a variable associated with an object. Functions include built-in functions or functions associated with a module. For example, enter the following into the interactive shell:

```
>>> len('Hello')
5
>>> 'Hello'.upper()
'HELLO'
>>> import math
>>> math.sqrt(25)
5.0
```

In this example, len() is a function and upper() is a string method. Methods are also considered attributes of the objects they're associated with. Note that a period doesn't necessarily mean you're dealing with a method instead of a function. The sqrt() function is associated with math, which is a module, not a class.

Iterable vs. Iterator

Python's for loops are versatile. The statement for i in range(3): will run a block of code three times. The range(3) call isn't just Python's way of telling a for loop, "repeat some code three times." Calling range(3) returns a range object, just like calling list('cat') returns a list object. Both of these objects are examples of *iterable objects* (or simply, *iterables*).

You use iterables in for loops. Enter the following into the interactive shell to see a for loop iterate over a range object and a list object:

```
>>> for i in range(3):
...     print(i) # body of the for loop
...
0
1
2
>>> for i in ['c', 'a', 't']:
...     print(i) # body of the for loop
...
c
a
t
```

Iterables also include all sequence types, such as range, list, tuple, and string objects, but also some container objects, such as dictionary, set, and file objects.

However, more is going on under the hood in these for loop examples. Behind the scenes, Python is calling the built-in iter() and next() functions for the for loop. When used in a for loop, iter*able* objects are passed to the built-in iter() function, which returns iter*ator* objects. Although the iterable object contains the items, the iterator object keeps track of which item is next to be used in a loop. On each iteration of the loop, the iterator object is passed to the built-in next() function to return the next item in the iterable. We can call the iter() and next() functions manually to directly see how for loops work. Enter the following into the interactive shell to perform the same instructions as the previous loop example:

```
>>> iterableObj = range(3)
>>> iterableObj
range(0, 3)
>>> iteratorObj = iter(iterableObj)
>>> i = next(iteratorObj)
>>> print(i) # body of the for loop
0
>>> i = next(iteratorObj)
>>> print(i) # body of the for loop
1
>>> i = next(iteratorObj)
>>> print(i) # body of the for loop
```

```
2
>>> i = next(iteratorObj)
Traceback (most recent call last):
  File "<stdin>", line 1, in <module>
❶ StopIteration
```

Notice that if you call next() after the last item in the iterable has been returned, Python raises a StopIteration exception ❶. Instead of crashing your programs with this error message, Python's for loops catch this exception to know when they should stop looping.

An iterator can only iterate over the items in an iterable once. This is similar to how you can only use open() and readlines() to read the contents of a file once before having to reopen the file to read its contents again. If you want to iterate over the iterable again, you must call iter() again to create another iterator object. You can create as many iterator objects as you want; each will independently track the next item it should return. Enter the following into the interactive shell to see how this works:

```
>>> iterableObj = list('cat')
>>> iterableObj
['c', 'a', 't']
>>> iteratorObj1 = iter(iterableObj)
>>> iteratorObj2 = iter(iterableObj)
>>> next(iteratorObj1)
'c'
>>> next(iteratorObj1)
'a'
>>> next(iteratorObj2)
'c'
```

Remember that iterable objects are passed as an argument to the iter() function, whereas the object returned from iter() calls is an iterator object. Iterator objects are passed to the next() function. When you create your own data types with class statements, you can implement the __iter__() and __next__() special methods to use your objects in for loops.

Syntax vs. Runtime vs. Semantic Errors

There are many ways to categorize bugs. But at a high level you could divide programming errors into three types: syntax errors, runtime errors, and semantic errors.

Syntax is the set of rules for the valid instructions in a given programming language. A *syntax error*, such as a missing parenthesis, a period instead of a comma, or some other typo will immediately generate a SyntaxError. Syntax errors are also known as *parsing errors*, which occur when the Python interpreter can't parse the text of the source code into valid instructions. In English, this error would be the equivalent of having incorrect grammar or a string of nonsense words like, "by uncontaminated cheese certainly it's." Computers require specific instructions and can't read the programmer's mind to determine what the program should do, so a program with a syntax error won't even run.

A *runtime error* is when a running program fails to perform some task, such as trying to open a file that doesn't exist or dividing a number by zero. In English, a runtime error is the equivalent of giving an impossible instruction like, "Draw a square with three sides." If a runtime error isn't addressed, the program will crash and display a traceback. But you can catch runtime errors using try-except statements that run error handling code. For example, enter the following into the interactive shell:

```
>>> slices = 8
>>> eaters = 0
>>> print('Each person eats', slices / eaters, 'slices.')
```

This code will display this traceback when you run it:

```
Traceback (most recent call last):
  File "<pyshell#4>", line 1, in <module>
    print('Each person eats', slices / eaters, 'slices.')
ZeroDivisionError: division by zero
```

It's helpful to remember that the line number the traceback mentions is only the point at which the Python interpreter detected an error. The true cause of the error might be on the previous line of code or even much earlier in the program.

Syntax errors in the source code are caught by the interpreter before the program runs, but syntax errors can also happen at runtime. The eval() function can take a string of Python code and run it, which might produce a SyntaxError at runtime. For example, eval('print("Hello, world)') is missing a closing double quote, which the program won't encounter until the code calls eval().

A *semantic error* (also called a *logical error*) is a more subtle bug. Semantic errors won't cause error messages or crashes, but the computer carries out instructions in a way the programmer didn't intend. In English, the equivalent of a semantic error would be telling the computer, "Buy a carton of milk from the store and if they have eggs, buy a dozen." The computer would then buy 13 cartons of milk because the store had eggs. For better or worse, computers do exactly what you tell them to. For example, enter the following into the interactive shell:

```
>>> print('The sum of 4 and 2 is', '4' + '2')
```

You would get the following output:

```
The sum of 4 and 2 is 42
```

Obviously, 42 isn't the answer. But notice that the program didn't crash. Because Python's + operator adds integer values and concatenates string values, mistakenly using the string values '4' and '2' instead of integers caused unintended behavior.

Parameters vs. Arguments

Parameters are the variable names between the parentheses in a def statement. *Arguments* are the values passed in a function call, which are then assigned to the parameters. For example, enter the following into the interactive shell:

```
❶ >>> def greeting(name, species):
...     print(name + ' is a ' + species)
...
❷ >>> greeting('Zophie', 'cat')
Zophie is a cat
```

In the def statement, name and species are parameters ❶. In the function call, 'Zophie' and 'cat' are arguments ❷. These two terms are often confused with each other. Remember that parameters and arguments are just other names for variables and values, respectively, when they are used in this context.

Type Coercion vs. Type Casting

You can convert an object of one type to an object of another type. For example, int('42') converts a string '42' to an integer 42. In actuality, the string object '42' isn't converted so much as the int() function creates a new integer object based on the original object. When conversion is done *explicitly* like this, we're *casting* the object, although programmers often still refer to this process as converting the object.

Python will often implicitly do a type conversion, such as when evaluating the expression 2 + 3.0 to 5.0. Values, such as the 2 and 3.0, are coerced to a common data type that the operator can work with. This conversion, which is done *implicitly*, is called type *coercion*.

Coercion can sometimes lead to surprising results. The Boolean True and False values in Python can be coerced to the integer values 1 and 0, respectively. Although you'd never write Booleans as those values in real-world code, this means that the expression True + False + True is the equivalent of 1 + 0 + 1 and evaluates to 2. After learning this, you might think that passing a list of Booleans to sum() would be a good way to count the number of True values in a list. But it turns out that calling the count() list method is faster.

Properties vs. Attributes

In many languages, the terms *property* and *attribute* are used synonymously, but in Python these words have distinct meanings. An attribute, explained in "Variable vs. Attribute" on page 124, is a name associated with an object. Attributes include the object's member variables and methods.

Other languages, such as Java, have getter and setter methods for classes. Instead of being able to directly assign an attribute a (potentially invalid) value, a program must call the setter method for that attribute.

The code inside the setter method can ensure that the member variable only has a valid value assigned to it. The getter method reads an attribute's value. If an attribute is named, say, `accountBalance`, the setter and getter methods are usually named `setAccountBalance()` and `getAccountBalance()`, respectively.

In Python, *properties* allow programmers to use getters and setters with much cleaner syntax. Chapter 17 explores Python properties in more detail.

Bytecode vs. Machine Code

Source code is compiled into a form of instructions called *machine code* that the CPU directly carries out. Machine code is composed of instructions from the CPU's *instruction set*, the computer's built-in set of commands. A compiled program composed of machine code is called a *binary*. A venerable language like C has compiler software that can compile C source code into binaries for almost every CPU available. But if a language such as Python wants to run on the same set of CPUs, a large amount of work would have to go into writing Python compilers for each of them.

There is another way of turning source code into machine-usable code. Instead of creating machine code that is carried out directly by CPU hardware, you could create *bytecode*. Also called *portable code* or *p-code*, bytecode is carried out by a software interpreter program instead of directly by the CPU. Python bytecode is composed of instructions from an instruction set, although no real-world hardware CPU carries out these instructions. Instead, the software interpreter executes the bytecode. Python bytecode is stored in the *.pyc* files you sometimes see alongside your *.py* source files. The CPython interpreter, which is written in C, can compile Python source code into Python bytecode and then carry out the instructions. (The same goes for the Java Virtual Machine [JVM] software, which carries out Java bytecode.) Because it's written in C, CPython has a Python interpreter and can be compiled for any CPU that C already has a compiler for.

The PyCon 2016 talk, "Playing with Python Bytecode" by Scott Sanderson and Joe Jevnik, is an excellent resource to learn more about this topic (*https://youtu.be/mxjv9KqzwjI*).

Script vs. Program, Scripting Language vs. Programming Language

The differences between a script and a program, or even a scripting language and a programming language, are vague and arbitrary. It's fair to say that all scripts are programs and all scripting languages are programming languages. But scripting languages are sometimes regarded as easier or "not real" programming languages.

One way to distinguish scripts from programs is by how the code executes. *Scripts* written in *scripting languages* are interpreted directly from the source code, whereas *programs* written in *programming languages* are compiled into binaries. But Python is commonly thought of as a scripting language, even though there is a compilation step to bytecode when a Python program is run. Meanwhile, Java isn't commonly thought of as a scripting language, even though it produces bytecode instead of machine code binaries, just like Python. Technically, *languages* aren't compiled or interpreted; rather, there are compiler or interpreter *implementations* of a language, and it's possible to create a compiler or interpreter for any language.

The differences can be argued but ultimately aren't very important. Scripting languages aren't necessarily less powerful, nor are compiled programming languages more difficult to work with.

Library vs. Framework vs. SDK vs. Engine vs. API

Using other people's code is a great time-saver. You can often find code to use packaged as libraries, frameworks, SDKs, engines, or APIs. The differences between these entities are subtle but important.

A *library* is a generic term for a collection of code made by a third party. A library can contain functions, classes, or other pieces of code for a developer to use. A Python library might take the form of a package, a set of packages, or even just a single module. Libraries are often specific to a particular language. The developer doesn't need to know how the library code works; they only need to know how to call or interface with the code in a library. A *standard library*, such as the Python standard library, is a code library that is assumed to be available to all implementations of a programming language.

A *framework* is a collection of code that operates with *inversion of control*; the developer creates functions that the framework will call as needed, as opposed to the developer's code calling functions in the framework. Inversion of control is often described as "don't call us, we'll call you." For example, writing code for a web app framework involves creating functions for the web pages that the framework will call when a web request comes in.

A *software development kit* (*SDK*) includes code libraries, documentation, and software tools to assist in creating applications for a particular operating system or platform. For example, the Android SDK and iOS SDK are used to create mobile apps for Android and iOS, respectively. The Java Development Kit (JDK) is an SDK for creating applications for the JVM.

An *engine* is a large, self-contained system that can be externally controlled by the developer's software. Developers usually call functions in an engine to perform a large, complex task. Examples of engines include game engines, physics engines, recommendation engines, database engines, chess engines, and search engines.

An *application programming interface* (*API*) is the public-facing interface for a library, SDK, framework, or engine. The API specifies how to call the functions or make requests of the library to access resources. The library creators will (hopefully) make documentation for the API available. Many popular social networks and websites make an HTTP API available for programs to access their services rather than a human with a web browser. Using these APIs allows you to write programs that can, for example, automatically post on Facebook or read Twitter timelines.

Summary

It's easy to program for years and still be unfamiliar with certain programming terms. But most major software applications are created by teams of software developers, not individuals. So being able to communicate unambiguously is important when you're working with a team.

This chapter explained that Python programs are made up of identifiers, variables, literals, keywords, and objects, and that all Python objects have a value, data type, and identity. Although every object has a data type, there are also several broad categories of types, such as container, sequence, mapping, set, built-in, and user-defined.

Some terms, like values, variables, and functions, have different names in specific contexts, such as items, parameters, arguments, and methods.

Several terms are also easy to confuse with each other. It's not a big deal to confuse some of these terms in day-to-day programming: for example, property versus attribute, block versus body, exception versus error, or the subtle differences between library, framework, SDK, engine, and API. Other misunderstandings won't make the code you write wrong but might make you look unprofessional: for example, statement and expression, function and method, and parameter and argument are commonly used interchangeably by beginners.

But other terms, such as iterable versus iterator, syntax error versus semantic error, and bytecode versus machine code, have distinct meanings that you should never confuse with each other unless you want to confuse your colleagues.

You'll still find that the use of terms varies from language to language and even programmer to programmer. You'll become more familiar with jargon with experience (and frequent web searches) in time.

Further Reading

The official Python glossary at *https://docs.python.org/3/glossary.html* lists short but helpful definitions the Python ecosystem uses. The official Python documentation at *https://docs.python.org/3/reference/datamodel.html* describes Python objects in greater detail.

Nina Zakharenko's PyCon 2016 talk, "Memory Management in Python—The Basics," at *https://youtu.be/F6u5rhUQ6dU*, explains many details about how Python's garbage collector works. The official Python documentation at *https://docs.python.org/3/library/gc.html* has more information about the garbage collector.

The Python mailing list discussion about making dictionaries ordered in Python 3.6 makes for good reading as well and is at *https://mail.python.org/pipermail/python-dev/2016-September/146327.html*.

8

COMMON PYTHON GOTCHAS

Although Python is my favorite programming language, it isn't without flaws. Every language has warts (some more than others), and Python is no exception. New Python programmers must learn to avoid some common "gotchas." Programmers learn this kind of knowledge randomly, from experience, but this chapter collects it in one place. Knowing the programming lore behind these gotchas can help you understand why Python behaves strangely sometimes.

This chapter explains how mutable objects, such as lists and dictionaries, can behave unexpectedly when you modify their contents. You'll learn how the sort() method doesn't sort items in an exact alphabetical order and how floating-point numbers can have rounding errors. The inequality operator != has unusual behavior when you chain them together. And you must use a trailing comma when you write tuples that contain a single item. This chapter informs you how to avoid these common gotchas.

Don't Add or Delete Items from a List While Looping Over It

Adding or deleting items from a list while looping (that is, *iterating*) over it with a for or while loop will most likely cause bugs. Consider this scenario: you want to iterate over a list of strings that describe items of clothing and ensure that there is an even number of socks by inserting a matching sock each time a sock is found in the list. The task seems straightforward: iterate over the list's strings, and when you find 'sock' in a string, such as 'red sock', append another 'red sock' string to the list.

But this code won't work. It gets caught in an infinite loop, and you'll have to press CTRL-C to interrupt it:

```
>>> clothes = ['skirt', 'red sock']
>>> for clothing in clothes:  # Iterate over the list.
...     if 'sock' in clothing:  # Find strings with 'sock'.
...         clothes.append(clothing)  # Add the sock's pair.
...         print('Added a sock:', clothing)  # Inform the user.
...
Added a sock: red sock
Added a sock: red sock
Added a sock: red sock
--snip--
Added a sock: red sock
Traceback (most recent call last):
  File "<stdin>", line 3, in <module>
KeyboardInterrupt
```

You'll find a visualization of the execution of this code at *https://autbor .com/addingloop/*.

The problem is that when you append 'red sock' to the clothes list, the list now has a new, third item that it must iterate over: ['skirt', 'red sock', 'red sock']. The for loop reaches the second 'red sock' on the next iteration, so it appends *another* 'red sock' string. This makes the list ['skirt', 'red sock', 'red sock', 'red sock'], giving the list another string for Python to iterate over. This will continue happening, as shown in Figure 8-1, which is why we see the never-ending stream of 'Added a sock.' messages. The loop only stops once the computer runs out of memory and crashes the Python program or until you interrupt it by pressing CTRL-C.

```
clothing
   ↓
['skirt', 'red sock']

              clothing
                 ↓
['skirt', 'red sock']

              clothing
                 ↓
['skirt', 'red sock', 'red sock']

                        clothing
                           ↓
['skirt', 'red sock', 'red sock']

                        clothing
                           ↓
['skirt', 'red sock', 'red sock', 'red sock']

                                  clothing
                                     ↓
['skirt', 'red sock', 'red sock', 'red sock']

                                  clothing
                                     ↓
['skirt', 'red sock', 'red sock', 'red sock', 'red sock']
   ⋮
```

Figure 8-1: On each iteration of the for loop, a new 'red sock' is appended to the list, which clothing refers to on the next iteration. This cycle repeats forever.

The takeaway is don't add items to a list while you're iterating over that list. Instead, use a separate list for the contents of the new, modified list, such as newClothes in this example:

```
>>> clothes = ['skirt', 'red sock', 'blue sock']
>>> newClothes = []
>>> for clothing in clothes:
...     if 'sock' in clothing:
...         print('Appending:', clothing)
...         newClothes.append(clothing)  # We change the newClothes list, not clothes.
...
Appending: red sock
Appending: blue sock
>>> print(newClothes)
['red sock', 'blue sock']
>>> clothes.extend(newClothes)  # Appends the items in newClothes to clothes.
>>> print(clothes)
['skirt', 'red sock', 'blue sock', 'red sock', 'blue sock']
```

A visualization of the execution of this code is at *https://autbor.com/addingloopfixed/*.

Our for loop iterated over the items in the clothes list but didn't modify clothes inside the loop. Instead, it changed a separate list, newClothes. Then, after the loop, we modify clothes by extending it with the contents of newClothes. You now have a clothes list with matching socks.

Similarly, you shouldn't delete items from a list while iterating over it. Consider code in which we want to remove any string that isn't 'hello' from a list. The naive approach is to iterate over the list, deleting the items that don't match 'hello':

```
>>> greetings = ['hello', 'hello', 'mello', 'yello', 'hello']
>>> for i, word in enumerate(greetings):
...     if word != 'hello':  # Remove everything that isn't 'hello'.
...         del greetings[i]
...
>>> print(greetings)
['hello', 'hello', 'yello', 'hello']
```

A visualization of the execution of this code is at *https://autbor.com/deletingloop/*.

It seems that 'yello' is left in the list. The reason is that when the for loop was examining index 2, it deleted 'mello' from the list. But this shifted all the remaining items in the list down one index, moving 'yello' from index 3 to index 2. The next iteration of the loop examines index 3, which is now the last 'hello', as in Figure 8-2. The 'yello' string slipped by unexamined! *Don't* remove items from a list while you're iterating over that list.

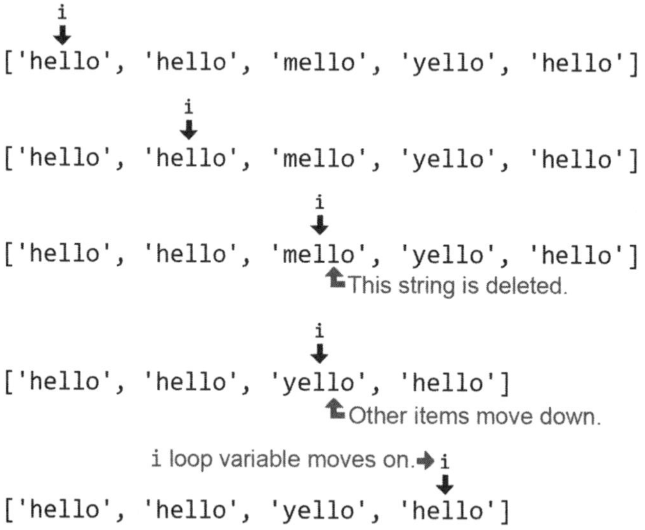

i
↓
['hello', 'hello', 'mello', 'yello', 'hello']

i
↓
['hello', 'hello', 'mello', 'yello', 'hello']

i
↓
['hello', 'hello', 'mello', 'yello', 'hello']
⬑This string is deleted.

i
↓
['hello', 'hello', 'yello', 'hello']
⬑Other items move down.

i loop variable moves on.➡ i
↓
['hello', 'hello', 'yello', 'hello']

Figure 8-2: When the loop removes 'mello', the items in the list shift down one index, causing i to skip over 'yello'.

Instead, create a new list that copies all the items except the ones you want to delete, and then replace the original list. For a bug-free equivalent of the previous example, enter the following code into the interactive shell.

```
>>> greetings = ['hello', 'hello', 'mello', 'yello', 'hello']
>>> newGreetings = []
>>> for word in greetings:
...     if word == 'hello':    # Copy everything that is 'hello'.
...         newGreetings.append(word)
...
>>> greetings = newGreetings  # Replace the original list.
>>> print(greetings)
['hello', 'hello', 'hello']
```

A visualization of the execution of this code is at *https://autbor.com/ deletingloopfixed/*.

Remember that because this code is just a simple loop that creates a list, you can replace it with a list comprehension. The list comprehension doesn't run faster or use less memory, but it's shorter to type without losing much readability. Enter the following into the interactive shell, which is equivalent to the code in the previous example:

```
>>> greetings = ['hello', 'hello', 'mello', 'yello', 'hello']
>>> greetings = [word for word in greetings if word == 'hello']
>>> print(greetings)
['hello', 'hello', 'hello']
```

Not only is the list comprehension more succinct, it also avoids the gotcha that occurs when changing a list while iterating over it.

REFERENCES, MEMORY USAGE, AND SYS.GETSIZEOF()

It might seem like creating a new list instead of modifying the original one wastes memory. But remember that, just as variables technically contain references to values instead of the actual values, lists also contain references to values. The newGreetings.append(word) line shown earlier isn't making a copy of the string in the word variable, just a copy of the reference to the string, which is much smaller.

You can see this by using the sys.getsizeof() function, which returns the number of bytes that the object passed to it takes up in memory. In this interactive shell example, we can see that the short string 'cat' takes up 52 bytes, whereas a longer string takes up 85 bytes:

```
>>> import sys
>>> sys.getsizeof('cat')
52
>>> sys.getsizeof('a much longer string than just "cat"')
85
```

(continued)

(In the Python version I use, the overhead for the string object takes up 49 bytes, whereas each actual character in the string takes up 1 byte.) But a list containing either of these strings takes up 72 bytes, no matter how long the string is:

```
>>> sys.getsizeof(['cat'])
72
>>> sys.getsizeof(['a much longer string than just "cat"'])
72
```

The reason is that a list technically doesn't contain the strings, but rather just a reference to the strings, and a reference is the same size no matter the size of the referred data. Code like newGreetings.append(word) isn't copying the string in word, but the reference to the string. If you want to find out how much memory an object, and all the objects it refers to, take up, Python core developer Raymond Hettinger has written a function for this, which you can access at *https://code.activestate.com/recipes/577504-compute-memory-footprint -of-an-object-and-its-cont/*.

So you shouldn't feel like you're wasting memory by creating a new list rather than modifying the original list while iterating over it. Even if your list-modifying code seemingly works, it can be the source of subtle bugs that take a long time to discover and fix. Wasting a programmer's time is far more expensive than wasting a computer's memory.

Although you shouldn't add or remove items from a list (or any iterable object) while iterating over it, it's fine to modify the list's contents. For example, say we have a list of numbers as strings: ['1', '2', '3', '4', '5']. We can convert this list of strings into a list of integers [1, 2, 3, 4, 5] while iterating over the list:

```
>>> numbers = ['1', '2', '3', '4', '5']
>>> for i, number in enumerate(numbers):
...     numbers[i] = int(number)
...
>>> numbers
[1, 2, 3, 4, 5]
```

A visualization of the execution of this code is at *https://autbor.com/ covertstringnumbers*. Modifying the items in the list is fine; it's changing the number of items in the list that is bug prone.

Another possible way to add or delete items in a list safely is by iterating backward from the end of the list to the beginning. This way, you can delete items from the list as you iterate over it, or add items to the list as long as you add them to the end of the list. For example, enter the following code, which removes even integers from the someInts list.

```
>>> someInts = [1, 7, 4, 5]
>>> for i in range(len(someInts)):
...
...         if someInts[i] % 2 == 0:
...             del someInts[i]
...
Traceback (most recent call last):
  File "<stdin>", line 2, in <module>
IndexError: list index out of range
>>> someInts = [1, 7, 4, 5]
>>> for i in range(len(someInts) - 1, -1, -1):
...         if someInts[i] % 2 == 0:
...             del someInts[i]
...
>>> someInts
[1, 7, 5]
```

This code works because none of the items that the loop will iterate over in the future ever have their index changed. But the repeated shifting up of values after the deleted value makes this technique inefficient for long lists. A visualization of the execution of this code is at *https://autbor.com/ iteratebackwards1*. You can see the difference between iterating forward and backward in Figure 8-3.

Iterating Forward

```
i
↓
[1, 7, 4, 5]
i starts at the front at index 0 and goes up to index 3.

    i
    ↓
[1, 7, 4, 5]
i reaches spam[2].

    i
    ↓
[1, 7, 5]
spam[2] is deleted, other values shift up.

        i
        ↓
[1, 7, 5]
i continues to index 3, which is now out of range.
```

Iterating Backward

```
        i
        ↓
[1, 7, 4, 5]
i starts at the end at index 3 and goes down to index 0.

        i
        ↓
[1, 7, 4, 5]
i reaches spam[2].

        i
        ↓
[1, 7, 5]
spam[2] is deleted, other values shift up.

i
↓
[1, 7, 5]
i continues to index 0, and finishes.
```

Figure 8-3: Removing even numbers from a list while iterating forward (left) and backward (right)

Similarly, you can add items to the end of the list as you iterate backward over it. Enter the following into the interactive shell, which appends a copy of any even integers in the someInts list to the end of the list:

```
>>> someInts = [1, 7, 4, 5]
>>> for i in range(len(someInts) - 1, -1, -1):
```

```
...        if someInts[i] % 2 == 0:
...            someInts.append(someInts[i])
...
>>> someInts
[1, 7, 4, 5, 4]
```

A visualization of the execution of this code is at *https://autbor.com/ iteratebackwards2*. By iterating backward, we can append items to or remove items from the list. But this can be tricky to do correctly because slight changes to this basic technique could end up introducing bugs. It's much simpler to create a new list rather than modifying the original list. As Python core developer Raymond Hettinger put it:

> **Q.** What are the best practices for modifying a list while looping over it?
> **A.** Don't.

Don't Copy Mutable Values Without copy.copy() and copy.deepcopy()

It's better to think of variables as labels or name tags that refer to objects rather than as boxes that contain objects. This mental model is especially useful when it comes to modifying *mutable* objects: objects such as lists, dictionaries, and sets whose value can mutate (that is, change). A common gotcha occurs when copying one variable that refers to a mutable object to another variable and thinking that the actual object is being copied. In Python, assignment statements never copy objects; they only copy the references to an object. (Python developer Ned Batchelder has a great PyCon 2015 talk on this idea titled, "Facts and Myths about Python Names and Values." Watch it at *https://youtu.be/_AEJHKGk9ns*.)

For example, enter the following code into the interactive shell, and note that even though we change the spam variable only, the cheese variable changes as well:

```
>>> spam = ['cat', 'dog', 'eel']
>>> cheese = spam
>>> spam
['cat', 'dog', 'eel']
>>> cheese
['cat', 'dog', 'eel']
>>> spam[2] = 'MOOSE'
>>> spam
['cat', 'dog', 'MOOSE']
>>> cheese
['cat', 'dog', 'MOOSE']
>>> id(cheese), id(spam)
2356896337288, 2356896337288
```

A visualization of the execution of this code is at *https://autbor.com/ listcopygotcha1*. If you think that cheese = spam copied the list object, you might be surprised that cheese seems to have changed even though we only modified spam. But assignment statements *never copy objects*, only references to objects. The assignment statement cheese = spam makes cheese *refer* to the same list object in the computer's memory as spam. It doesn't *duplicate* the list object. This is why changing spam also changes cheese: both variables refer to the same list object.

The same principle applies to mutable objects passed to a function call. Enter the following into the interactive shell, and note that the global variable spam and the local parameter (remember, parameters are variables defined in the function's def statement) theList both refer to the same object:

```
>>> def printIdOfParam(theList):
...     print(id(theList))
...
>>> eggs = ['cat', 'dog', 'eel']
>>> print(id(eggs))
2356893256136
>>> printIdOfParam(eggs)
2356893256136
```

A visualization of the execution of this code is at *https://autbor.com/ listcopygotcha2*. Notice that the identities returned by id() for eggs and theList are the same, meaning these variables refer to the same list object. The eggs variable's list object wasn't copied to theList; rather, the reference was copied, which is why both variables refer to the same list. A reference is only a few bytes in size, but imagine if Python copied the entire list instead of just the reference. If eggs contained a billion items instead of just three, passing it to the printIdOfParam() function would require copying this giant list. This would eat up gigabytes of memory just to do a simple function call! That's why Python assignment only copies references and never copies objects.

One way to prevent this gotcha is to make a copy of the list object (not just the reference) with the copy.copy() function. Enter the following into the interactive shell:

```
>>> import copy
>>> bacon = [2, 4, 8, 16]
>>> ham = copy.copy(bacon)
>>> id(bacon), id(ham)
(2356896337352, 2356896337480)
>>> bacon[0] = 'CHANGED'
>>> bacon
['CHANGED', 4, 8, 16]
>>> ham
[2, 4, 8, 16]
>>> id(bacon), id(ham)
(2356896337352, 2356896337480)
```

A visualization of the execution of this code is at *https://autbor.com/ copycopy1*. The ham variable refers to a copied list object rather than the original list object referred to by bacon, so it doesn't suffer from this gotcha.

But just as variables are like labels or name tags rather than boxes that contain objects, lists also contain labels or name tags that refer to objects rather than the actual objects. If your list contains other lists, copy.copy() only copies the references to these inner lists. Enter the following into the interactive shell to see this problem:

```
>>> import copy
>>> bacon = [[1, 2], [3, 4]]
>>> ham = copy.copy(bacon)
>>> id(bacon), id(ham)
(2356896466248, 2356896375368)
>>> bacon.append('APPENDED')
>>> bacon
[[1, 2], [3, 4], 'APPENDED']
>>> ham
[[1, 2], [3, 4]]
>>> bacon[0][0] = 'CHANGED'
>>> bacon
[['CHANGED', 2], [3, 4], 'APPENDED']
>>> ham
[['CHANGED', 2], [3, 4]]
>>> id(bacon[0]), id(ham[0])
(2356896337480, 2356896337480)
```

A visualization of the execution of this code is at *https://autbor.com/copycopy2*. Although bacon and ham are two different list objects, they refer to the same [1, 2] and [3, 4] inner lists, so changes to these inner lists get reflected in both variables, even though we used copy.copy(). The solution is to use copy.deepcopy(), which will make copies of any list objects inside the list object being copied (and any list objects in those list objects, and so on). Enter the following into the interactive shell:

```
>>> import copy
>>> bacon = [[1, 2], [3, 4]]
>>> ham = copy.deepcopy(bacon)
>>> id(bacon[0]), id(ham[0])
(2356896337352, 2356896466184)
>>> bacon[0][0] = 'CHANGED'
>>> bacon
[['CHANGED', 2], [3, 4]]
>>> ham
[[1, 2], [3, 4]]
```

A visualization of the execution of this code is at *https://autbor.com/copydeepcopy*. Although copy.deepcopy() is slightly slower than copy.copy(), it's safer to use if you don't know whether the list being copied contains other lists (or other mutable objects like dictionaries or sets). My general advice is to always use copy.deepcopy(): it might prevent subtle bugs, and the slowdown in your code probably won't be noticeable.

Don't Use Mutable Values for Default Arguments

Python allows you to set *default arguments* for parameters in the functions you define. If a user doesn't explicitly set a parameter, the function will execute using the default argument. This is useful when most calls to the function use the same argument, because default arguments make the parameter optional. For example, passing None for the split() method makes it split on whitespace characters, but None is also the default argument: calling 'cat dog'.split() does the same thing as calling 'cat dog'.split(None). The function uses the default argument for the parameter's argument unless the caller passes one in.

But you should never set a *mutable* object, such as a list or dictionary, as a default argument. To see how this causes bugs, look at the following example, which defines an addIngredient() function that adds an ingredient string to a list that represents a sandwich. Because the first and last items of this list are often 'bread', the mutable list ['bread', 'bread'] is used as a default argument:

```
>>> def addIngredient(ingredient, sandwich=['bread', 'bread']):
...     sandwich.insert(1, ingredient)
...     return sandwich
...
>>> mySandwich = addIngredient('avocado')
>>> mySandwich
['bread', 'avocado', 'bread']
```

But using a mutable object, such as a list like ['bread', 'bread'], for the default argument has a subtle problem: the list is created when the function's def statement executes, not each time the function is called. This means that only one ['bread', 'bread'] list object gets created, because we only *define* the addIngredient() function once. But each function *call* to addIngredient() will be reusing this list. This leads to unexpected behavior, like the following:

```
>>> mySandwich = addIngredient('avocado')
>>> mySandwich
['bread', 'avocado', 'bread']
>>> anotherSandwich = addIngredient('lettuce')
>>> anotherSandwich
['bread', 'lettuce', 'avocado', 'bread']
```

Because addIngredient('lettuce') ends up using the same default argument list as the previous calls, which already had 'avocado' added to it, instead of ['bread', 'lettuce', 'bread'] the function returns ['bread', 'lettuce', 'avocado', 'bread']. The 'avocado' string appears again because the list for the sandwich parameter is the same as the last function call. Only one ['bread', 'bread'] list was created, because the function's def statement only executes once, not each time the function is called. A visualization of the execution of this code is at *https://autbor.com/sandwich*.

If you need to use a list or dictionary as a default argument, the pythonic solution is to set the default argument to None. Then have code that checks for this and supplies a new list or dictionary whenever the function is called. This ensures that the function creates a new mutable object *each time* the function is called instead of *just once* when the function is defined, such as in the following example:

```
>>> def addIngredient(ingredient, sandwich=None):
...     if sandwich is None:
...         sandwich = ['bread', 'bread']
...     sandwich.insert(1, ingredient)
...     return sandwich
...
>>> firstSandwich = addIngredient('cranberries')
>>> firstSandwich
['bread', 'cranberries', 'bread']
>>> secondSandwich = addIngredient('lettuce')
>>> secondSandwich
['bread', 'lettuce', 'bread']
>>> id(firstSandwich) == id(secondSandwich)
❶ False
```

Notice that firstSandwich and secondSandwich don't share the same list reference ❶ because sandwich = ['bread', 'bread'] creates a new list object each time addIngredient() is called, not just once when addIngredient() is defined.

Mutable data types include lists, dictionaries, sets, and objects made from the class statement. *Don't* put objects of these types as default arguments in a def statement.

Don't Build Strings with String Concatenation

In Python, strings are *immutable* objects. This means that string values can't change, and any code that seems to modify the string is actually creating a new string object. For example, each of the following operations changes the content of the spam variable, not by changing the string value, but by replacing it with a new string value that has a new identity:

```
>>> spam = 'Hello'
>>> id(spam), spam
(38330864, 'Hello')
>>> spam = spam + ' world!'
>>> id(spam), spam
(38329712, 'Hello world!')
>>> spam = spam.upper()
>>> id(spam), spam
(38329648, 'HELLO WORLD!')
>>> spam = 'Hi'
>>> id(spam), spam
(38395568, 'Hi')
>>> spam = f'{spam} world!'
```

```
>>> id(spam), spam
(38330864, 'Hi world!')
```

Notice that each call to id(spam) returns a different identity, because the string object in spam isn't being changed: it's being replaced by a whole new string object with a different identity. Creating new strings by using f-strings, the format() string method, or the %s format specifiers also creates new string objects, just like string concatenation. Normally, this technical detail doesn't matter. Python is a high-level language that handles many of these details for you so you can focus on creating your program.

But building a string through a large number of string concatenations can slow down your programs. Each iteration of the loop creates a new string object and discards the old string object: in code, this looks like concatenations inside a for or while loop, as in the following:

```
>>> finalString = ''
>>> for i in range(100000):
...     finalString += 'spam '
...
>>> finalString
spam spam spam spam spam spam spam spam spam spam spam spam --snip--
```

Because the finalString += 'spam ' happens 100,000 times inside the loop, Python is performing 100,000 string concatenations. The CPU has to create these intermediate string values by concatenating the current finalString with 'spam ', put them into memory, and then almost immediately discard them on the next iteration. This is a lot of wasted effort, because we only care about the final string.

The pythonic way to build strings is to append the smaller strings to a list and then join the list together into one string. This method still creates 100,000 string objects, but it only performs one string concatenation, when it calls join(). For example, the following code produces the equivalent finalString but without the intermediate string concatenations:

```
>>> finalString = []
>>> for i in range(100000):
...     finalString.append('spam ')
...
>>> finalString = ''.join(finalString)
>>> finalString
spam spam spam spam spam spam spam spam spam spam spam spam --snip--
```

When I measure the runtime of these two pieces of code on my machine, the list appending approach is *10 times faster* than the string concatenation approach. (Chapter 13 describes how to measure how fast your programs run.) This difference becomes greater the more iterations the for loop makes. But when you change range(100000) to range(100), although concatenation remains slower than list appending, the speed difference is negligible. You don't need to obsessively avoid string concatenation, f-strings, the format() string method, or %s format specifiers in every case. The speed only significantly improves when you're performing large numbers of string concatenations.

Python frees you from having to think about many underlying details. This allows programmers to write software quickly, and as mentioned earlier, programmer time is more valuable than CPU time. But there are cases when it's good to understand details, such as the difference between immutable strings and mutable lists, to avoid tripping on a gotcha, like building strings through concatenation.

Don't Expect sort() to Sort Alphabetically

Understanding sorting algorithms—algorithms that systematically arrange values by some established order—is an important foundation for a computer science education. But this isn't a computer science book; we don't need to know these algorithms, because we can just call Python's sort() method. However, you'll notice that sort() has some odd sorting behavior that puts a capital *Z* before a lowercase *a*:

```
>>> letters = ['z', 'A', 'a', 'Z']
>>> letters.sort()
>>> letters
['A', 'Z', 'a', 'z']
```

The American Standard Code for Information Interchange (ASCII, pronounced "ask-ee") is a mapping between numeric codes (called *code points* or *ordinals*) and text characters. The sort() method uses *ASCII-betical* sorting (a general term meaning sorted by ordinal number) rather than alphabetical sorting. In the ASCII system, *A* is represented by code point 65, *B* by 66, and so on, up to *Z* by 90. The lowercase *a* is represented by code point 97, *b* by 98, and so on, up to *z* by 122. When sorting by ASCII, uppercase *Z* (code point 90) comes before lowercase *a* (code point 97).

Although it was almost universal in Western computing prior to and throughout the 1990s, ASCII is an American standard only: there's a code point for the dollar sign, $ (code point 36), but there is no code point for the British pound sign, £. ASCII has largely been replaced by Unicode, because Unicode contains all of ASCII's code points and more than 100,000 other code points.

You can get the code point, or ordinal, of a character by passing it to the ord() function. You can do the reverse by passing an ordinal integer to the chr() function, which returns a string of the character. For example, enter the following into the interactive shell:

```
>>> ord('a')
97
>>> chr(97)
'a'
```

If you want to make an alphabetical sort, pass the str.lower method to the key parameter. This sorts the list as if the values had the lower() string method called on them:

```
>>> letters = ['z', 'A', 'a', 'Z']
>>> letters.sort(key=str.lower)
>>> letters
['A', 'a', 'z', 'Z']
```

Note that the actual strings in the list aren't converted to lowercase; they're only sorted as if they were. Ned Batchelder provides more information about Unicode and code points in his talk "Pragmatic Unicode, or, How Do I Stop the Pain?" at *https://nedbatchelder.com/text/unipain.html*.

Incidentally, the sorting algorithm that Python's sort() method uses is Timsort, which was designed by Python core developer and "Zen of Python" author Tim Peters. It's a hybrid of the merge sort and insertion sort algorithms, and is described at *https://en.wikipedia.org/wiki/Timsort*.

Don't Assume Floating-Point Numbers Are Perfectly Accurate

Computers can only store the digits of the binary number system, which are 1 and 0. To represent the decimal numbers we're familiar with, we need to translate a number like 3.14 into a series of binary ones and zeros. Computers do this according to the IEEE 754 standard, published by the Institute of Electrical and Electronics Engineers (IEEE, pronounced "eye-triple-ee"). For simplicity, these details are hidden from the programmer, allowing you to type numbers with decimal points and ignore the decimal-to-binary conversion process:

```
>>> 0.3
0.3
```

Although the details of specific cases are beyond the scope of this book, the IEEE 754 representation of a floating-point number won't always exactly match the decimal number. One well-known example is 0.1:

```
>>> 0.1 + 0.1 + 0.1
0.30000000000000004
>>> 0.3 == (0.1 + 0.1 + 0.1)
False
```

This bizarre, slightly inaccurate sum is the result of *rounding errors* caused by how computers represent and process floating-point numbers. This isn't a *Python* gotcha; the IEEE 754 standard is a *hardware* standard implemented directly into a CPU's floating-point circuits. You'll get the same results in C++, JavaScript, and every other language that runs on a CPU that uses IEEE 754 (which is effectively every CPU in the world).

The IEEE 754 standard, again for technical reasons beyond the scope of this book, also cannot represent all whole number values greater than 2^{53}. For example, 2^{53} and $2^{53} + 1$, as float values, both round to 9007199254740992.0:

```
>>> float(2**53) == float(2**53) + 1
True
```

As long as you use the floating-point data type, there's no workaround for these rounding errors. But don't worry. Unless you're writing software for a bank, a nuclear reactor, or a bank's nuclear reactor, rounding errors are small enough that they'll likely not be an important issue for your program. Often, you can resolve them by using integers with smaller denominations: for example, 133 cents instead of 1.33 dollars or 200 milliseconds instead of 0.2 seconds. This way, 10 + 10 + 10 adds up to 30 cents or milliseconds rather than 0.1 + 0.1 + 0.1 adding up to 0.30000000000000004 dollars or seconds.

But if you need exact precision, say for scientific or financial calculations, use Python's built-in decimal module, which is documented at *https:// docs.python.org/3/library/decimal.html*. Although they're slower, Decimal objects are precise replacements for float values. For example, decimal.Decimal('0.1') creates an object that represents the exact number 0.1 without the imprecision that a 0.1 float value would have.

Passing the float value 0.1 to decimal.Decimal() creates a Decimal object that has the same imprecision as a float value, which is why the resulting Decimal object isn't exactly Decimal('0.1'). Instead, pass a string of the float value to decimal.Decimal(). To illustrate this point, enter the following into the interactive shell:

```
>>> import decimal
>>> d = decimal.Decimal(0.1)
>>> d
Decimal('0.1000000000000000055511151231257827021181583404541015625')
>>> d = decimal.Decimal('0.1')
>>> d
Decimal('0.1')
>>> d + d + d
Decimal('0.3')
```

Integers don't have rounding errors, so it's always safe to pass them to decimal.Decimal(). Enter the following into the interactive shell:

```
>>> 10 + d
Decimal('10.1')
>>> d * 3
Decimal('0.3')
>>> 1 - d
Decimal('0.9')
>>> d + 0.1
Traceback (most recent call last):
  File "<stdin>", line 1, in <module>
TypeError: unsupported operand type(s) for +: 'decimal.Decimal' and 'float'
```

But Decimal objects don't have unlimited precision; they simply have a predictable, well-established level of precision. For example, consider the following operations:

```
>>> import decimal
>>> d = decimal.Decimal(1) / 3
>>> d
Decimal('0.3333333333333333333333333333')
>>> d * 3
Decimal('0.9999999999999999999999999999')
>>> (d * 3) == 1 # d is not exactly 1/3
False
```

The expression `decimal.Decimal(1) / 3` evaluates to a value that isn't exactly one-third. But by default, it'll be precise to 28 significant digits. You can find out how many significant digits the `decimal` module uses by accessing the `decimal.getcontext().prec` attribute. (Technically, `prec` is an attribute of the `Context` object returned by `getcontext()`, but it's convenient to put it on one line.) You can change this attribute so that all `Decimal` objects created afterward use this new level of precision. The following interactive shell example lowers the precision from the original 28 significant digits to 2:

```
>>> import decimal
>>> decimal.getcontext().prec
28
>>> decimal.getcontext().prec = 2
>>> decimal.Decimal(1) / 3
Decimal('0.33')
```

The `decimal` module provides you with fine control over how numbers interact with each other. The `decimal` module is documented in full at *https://docs.python.org/3/library/decimal.html*.

Don't Chain Inequality != Operators

Chaining comparison operators like `18 < age < 35` or chaining assignment operators like `six = halfDozen = 6` are handy shortcuts for `(18 < age) and (age < 35)` and `six = 6; halfDozen = 6`, respectively.

But don't chain the `!=` comparison operator. You might think the following code checks whether all three variables have different values from each other, because the following expression evaluates to `True`:

```
>>> a = 'cat'
>>> b = 'dog'
>>> c = 'moose'
>>> a != b != c
True
```

But this chain is actually equivalent to (a != b) and (b != c). This means that a could still be the same as c and the a != b != c expression would still be True:

```
>>> a = 'cat'
>>> b = 'dog'
>>> c = 'cat'
>>> a != b != c
True
```

This bug is subtle and the code is misleading, so it's best to avoid using chained != operators altogether.

Don't Forget the Comma in Single-Item Tuples

When writing tuple values in your code, keep in mind that you'll still need a trailing comma even if the tuple only contains a single item. Although the value (42,) is a tuple that contains the integer 42, the value (42) is simply the integer 42. The parentheses in (42) are similar to those used in the expression (20 + 1) * 2, which evaluates to the integer value 42. Forgetting the comma can lead to this:

```
>>> spam = ('cat', 'dog', 'moose')
>>> spam[0]
'cat'
>>> spam = ('cat')
❶ >>> spam[0]
'c'
❷ >>> spam = ('cat', )
>>> spam[0]
'cat'
```

Without a comma, ('cat') evaluates to the string value, which is why spam[0] evaluates to the first character of the string, 'c' ❶. The trailing comma is required for the parentheses to be recognized as a tuple value ❷. In Python, the commas make a tuple more than the parentheses.

Summary

Miscommunication happens in every language, even in programming languages. Python has a few gotchas that can trap the unwary. Even if they rarely come up, it's best to know about them so you can quickly recognize and debug the problems they can cause.

Although it's possible to add or remove items from a list while iterating over that list, it's a potential source of bugs. It's much safer to iterate over a copy of the list and then make changes to the original. When you do make copies of a list (or any other mutable object), remember that assignment statements copy only the reference to the object, not the actual object. You can use the `copy.deepcopy()` function to make copies of the object (and copies of any objects it references).

You shouldn't use mutable objects in `def` statements for default arguments, because they're created once when the `def` statement is run rather than each time the function is called. A better idea is to make the default argument `None`, and then add code that checks for `None` and creates a mutable object when the function is called.

A subtle gotcha is the string concatenation of several smaller strings with the + operator in a loop. For small numbers of iteration, this syntax is fine. But under the hood, Python is constantly creating and destroying string objects on each iteration. A better approach is to append the smaller strings into a list and then call the `join()` operator to create the final string.

The `sort()` method sorts by numeric code points, which isn't the same as alphabetical order: uppercase Z is sorted before lowercase a. To fix this issue, you can call `sort(key=str.lower)`.

Floating-point numbers have slight rounding errors as a side effect of how they represent numbers. For most programs, this isn't important. But if it does matter for your program, you can use Python's `decimal` module.

Never chain together `!=` operators, because expressions like `'cat' != 'dog' != 'cat'` will, confusingly, evaluate to `True`.

Although this chapter described the Python gotchas that you're most likely to encounter, they don't occur daily in most real-world code. Python does a great job of minimizing the surprises you might find in your programs. In the next chapter, we'll cover some gotchas that are even rarer and downright bizarre. It's almost impossible that you'll ever encounter these Python language oddities if you aren't searching for them, but it'll be fun to explore the reasons they exist.

9

ESOTERIC PYTHON ODDITIES

The systems of rules that define a programming language are complicated and can lead to code that, although not wrong, is quite odd and unexpected. This chapter dives into the more obscure Python language oddities. You're unlikely to actually run into these cases in real-world coding, but they're interesting uses of the Python syntax (or abuses of it, depending on your perspective).

By studying the examples in this chapter, you'll get a better idea of how Python works under the hood. Let's have a little fun and explore some esoteric gotchas.

Why 256 Is 256 but 257 Is Not 257

The == operator compares two objects for equal value, but the is operator compares them for equal identity. Although the integer value 42 and the float value 42.0 have the same value, they're two different objects held in separate places in the computer's memory. You can confirm this by checking their different IDs using the id() function:

```
>>> a = 42
>>> b = 42.0
>>> a == b
True
>>> a is b
False
>>> id(a), id(b)
(140718571382896, 2526629638888)
```

When Python creates a new integer object and stores it in memory, that object creation takes very little time. As a tiny optimization, CPython (the Python interpreter available for download at *https://python.org*) creates integer objects for -5 to 256 at the start of every program. These integers are called *preallocated integers*, and CPython automatically creates objects for them because they're fairly common: a program is more likely to use the integer 0 or 2 than, say, 1729. When creating a new integer object in memory, CPython first checks whether it's between -5 and 256. If so, CPython saves time by simply returning the existing integer object instead of creating a new one. This behavior also saves memory by not storing duplicate small integers, as illustrated in Figure 9-1.

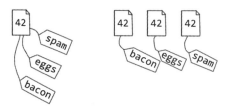

Figure 9-1: Python saves memory by using multiple references to a single integer object (left) instead of separate, duplicate integer objects for each reference (right).

Because of this optimization, certain contrived situations can produce bizarre results. To see an example of one, enter the following into the interactive shell:

```
>>> a = 256
>>> b = 256
❶ >>> a is b
True
>>> c = 257
>>> d = 257
❷ >>> c is d
False
```

All 256 objects are really the same object, so the is operator for a and b returns True ❶. But Python created separate 257 objects for c and d, which is why the is operator returns False ❷.

The expression 257 is 257 evaluates to True, but CPython reuses the integer object made for identical literals in the same statement:

```
>>> 257 is 257
True
```

Of course, real-world programs usually only use an integer's value, not its identity. They would never use the is operator to compare integers, floats, strings, bools, or values of other simple data types. One exception occurs when you use is None instead of == None, as explained in "Use is to Compare with None Instead of ==" on page 96. Otherwise, you'll rarely run into this problem.

String Interning

Similarly, Python reuses objects to represent identical string literals in your code rather than making separate copies of the same string. To see this in practice, enter the following into the interactive shell:

```
>>> spam = 'cat'
>>> eggs = 'cat'
>>> spam is eggs
True
>>> id(spam), id(eggs)
(1285806577904, 1285806577904)
```

Python notices that the 'cat' string literal assigned to eggs is the same as the 'cat' string literal assigned to spam; so instead of making a second, redundant string object, it just assigns eggs a reference to the same string object that spam uses. This explains why the IDs of their strings are the same.

This optimization is called *string interning*, and like the preallocated integers, it's nothing more than a CPython implementation detail. You should never write code that relies on it. Also, this optimization won't catch every possible identical string. Trying to identify every instance in which you can use an optimization often takes up more time than the optimization would save. For example, try creating the 'cat' string from 'c' and 'at' in the interactive shell; you'll notice that CPython creates the final 'cat' string as a new string object rather than reusing the string object made for spam:

```
>>> bacon = 'c'
>>> bacon += 'at'
>>> spam is bacon
False
>>> id(spam), id(bacon)
(1285806577904, 1285808207384)
```

String interning is an optimization technique that interpreters and compilers use for many different languages. You'll find further details at *https://en.wikipedia.org/wiki/String_interning.*

Python's Fake Increment and Decrement Operators

In Python, you can increase the value of a variable by 1 or reduce it by 1 using the augmented assignment operators. The code spam += 1 and spam -= 1 increments and decrements the numeric values in spam by 1, respectively.

Other languages, such as C++ and JavaScript, have the ++ and -- operators for incrementing and decrementing. (The name "C++" itself reflects this; it's a tongue-in-cheek joke that indicates it's an enhanced form of the C language.) Code in C++ and JavaScript could have operations like ++spam or spam++. Python wisely doesn't include these operators because they're notoriously susceptible to subtle bugs (as discussed at *https://softwareengineering .stackexchange.com/q/59880*).

But it's perfectly legal to have the following Python code:

```
>>> spam = --spam
>>> spam
42
```

The first detail you should notice is that the ++ and -- "operators" in Python don't actually increment or decrement the value in spam. Rather, the leading - is Python's unary negation operator. It allows you to write code like this:

```
>>> spam = 42
>>> -spam
-42
```

It's legal to have multiple unary negative operators in front of a value. Using two of them gives you the negative of the negative of the value, which for integer values just evaluates to the original value:

```
>>> spam = 42
>>> -(-spam)
42
```

This is a very silly operation to perform, and you likely won't ever see a unary negation operator used twice in real-world code. (But if you did, it's probably because the programmer learned to program in another language and has just written buggy Python code!)

There is also a + unary operator. It evaluates an integer value to the same sign as the original value, which is to say, it does absolutely nothing:

```
>>> spam = 42
>>> +spam
42
>>> spam = -42
```

```
>>> +spam
-42
```

Writing +42 (or ++42) seems just as silly as --42, so why does Python even have this unary operator? It exists only to complement the - operator if you need to overload these operators for your own classes. (That's a lot of terms you might not be familiar with! You'll learn more about operator overloading in Chapter 17.)

The + and - unary operators are only valid when in front of a Python value, not after it. Although spam++ and spam-- might be legal code in C++ or JavaScript, they produce syntax errors in Python:

```
>>> spam++
  File "<stdin>", line 1
    spam++
         ^
SyntaxError: invalid syntax
```

Python doesn't have increment and decrement operators. A quirk of the language syntax merely makes it seem like it does.

All of Nothing

The all() built-in function accepts a sequence value, such as a list, and returns True if all the values in that sequence are "truthy." It returns False if one or more values are "falsey." You can think of the function call all([False, True, True]) as equivalent to the expression False and True and True.

You can use all() in conjunction with list comprehensions to first create a list of Boolean values based on another list and then evaluate their collective value. For example, enter the following into the interactive shell:

```
>>> spam = [67, 39, 20, 55, 13, 45, 44]
>>> [i > 42 for i in spam]
[True, False, False, True, False, True, True]
>>> all([i > 42 for i in spam])
False
>>> eggs = [43, 44, 45, 46]
>>> all([i > 42 for i in eggs])
True
```

The all() utility returns True if all numbers in spam or eggs are greater than 42.

But if you pass an empty sequence to all(), it always returns True. Enter the following into the interactive shell:

```
>>> all([])
True
```

It's best to think of all([]) as evaluating the claim "none of the items in this list are falsey" instead of "all the items in this list are truthy." Otherwise, you might get some odd results. For instance, enter the following into the interactive shell:

```
>>> spam = []
>>> all([i > 42 for i in spam])
True
>>> all([i < 42 for i in spam])
True
>>> all([i == 42 for i in spam])
True
```

This code seems to be showing that not only are all the values in spam (an empty list) greater than 42, but they're also less than 42 and exactly equal to 42! This seems logically impossible. But remember that each of these three list comprehensions evaluates to the empty list, which is why none of the items in them are falsey and the all() function returns True.

Boolean Values Are Integer Values

Just as Python considers the float value 42.0 to be equal to the integer value 42, it considers the Boolean values True and False to be equivalent to 1 and 0, respectively. In Python, the bool data type is a subclass of the int data type. (We'll cover classes and subclasses in Chapter 16.) You can use int() to convert Boolean values to integers:

```
>>> int(False)
0
>>> int(True)
1
>>> True == 1
True
>>> False == 0
True
```

You can also use isinstance() to confirm that a Boolean value is considered a type of integer:

```
>>> isinstance(True, bool)
True
>>> isinstance(True, int)
True
```

The value True is of the bool data type. But because bool is a subclass of int, True is also an int. This means you can use True and False in almost any place you can use integers. This can lead to some bizarre code:

```
>>> True + False + True + True   # Same as 1 + 0 + 1 + 1
3
>>> -True            # Same as -1.
-1
```

```
>>> 42 * True        # Same as 42 * 1 mathematical multiplication.
42
>>> 'hello' * False  # Same as 'hello' * 0 string replication.
' '
>>> 'hello'[False]   # Same as 'hello'[0]
'h'
>>> 'hello'[True]    # Same as 'hello'[1]
'e'
>>> 'hello'[-True]   # Same as 'hello'[-1]
'o'
```

Of course, just because you can use bool values as numbers doesn't mean you should. The previous examples are all unreadable and should never be used in real-world code. Originally, Python didn't have a bool data type. It didn't add Booleans until Python 2.3, at which point it made bool a subclass of int to ease the implementation. You can read the history of the bool data type in PEP 285 at *https://www.python.org/dev/peps/pep-0285/*.

Incidentally, True and False were only made keywords in Python 3. This means that in Python 2, it was possible to use True and False as variable names, leading to seemingly paradoxical code like this:

```
Python 2.7.14 (v2.7.14:84471935ed, Sep 16 2017, 20:25:58) [MSC v.1500 64 bit
(AMD64)] on win32
Type "help", "copyright", "credits" or "license" for more information.
>>> True is False
False
>>> True = False
>>> True is False
True
```

Fortunately, this sort of confusing code isn't possible in Python 3, which will raise a syntax error if you try to use the keywords True or False as variable names.

Chaining Multiple Kinds of Operators

Chaining different kinds of operators in the same expression can produce unexpected bugs. For example, this (admittedly unrealistic) example uses the == and in operators in a single expression:

```
>>> False == False in [False]
True
```

This True result is surprising, because you would expect it to evaluate as either:

- (False == False) in [False], which is False.
- False == (False in [False]), which is also False.

But False == False in [False] isn't equivalent to either of these expressions. Rather, it's equivalent to (False == False) and (False in [False]), just as 42 < spam < 99 is equivalent to (42 < spam) and (spam < 99). This expression evaluates according to the following diagram:

```
(False == False) and (False in [False])
                  ↓
    (True) and (False in [False])
                  ↓
        (True) and (True)
                  ↓
              True
```

The False == False in [False] expression is a fun Python riddle, but it's unlikely to come up in any real-world code.

Python's Antigravity Feature

To enable Python's antigravity feature, enter the following into the interactive shell:

```
>>> import antigravity
```

This line is a fun Easter egg that opens the web browser to a classic XKCD comic strip about Python at *https://xkcd.com/353/.* It might surprise you that Python can open your web browser, but this is a built-in feature the webbrowser module provides. Python's webbrowser module has an open() function that finds your operating system's default web browser and opens a browser window to a specific URL. Enter the following into the interactive shell:

```
>>> import webbrowser
>>> webbrowser.open('https://xkcd.com/353/')
```

The webbrowser module is limited, but it can be useful for directing the user to further information on the internet.

Summary

It's easy to forget that computers and programming languages are designed by humans and have their own limitations. So much software is built on top of and relies upon the creations of language designers and hardware engineers. They work incredibly hard to make sure that if you have a bug in your program, it's because your program is faulty, not the interpreter software or CPU hardware running it. We can end up taking these tools for granted.

But this is why there's value in learning the odd nooks and crannies of computers and software. When your code raises errors or crashes (or even just acts weirdly and makes you think, "that's odd"), you'll need to understand the common gotchas to debug these problems.

You almost certainly won't run into any of the issues brought up in this chapter, but being aware of these small details is what will make you an experienced Python programmer.

10

WRITING EFFECTIVE FUNCTIONS

Functions are like mini programs within our programs that allow us to break code into smaller units. This spares us from having to write duplicate code, which can introduce bugs. But writing effective functions requires making many decisions about naming, size, parameters, and complexity.

This chapter explores the different ways we can write functions and the benefits and drawbacks of different trade-offs. We'll delve into how to trade off between small and large functions, how the number of parameters affects the function's complexity, and how to write functions with variable numbers of arguments using the * and ** operators. We'll also explore the functional programming paradigm and the benefits of writing functions according to this paradigm.

Function Names

Function names should follow the same convention we use for identifiers in general, as described in Chapter 4. But they should usually include a verb, because functions typically perform some action. You might also include a noun to describe the thing being acted on. For example, the names refreshConnection(), setPassword(), and extract_version() clarify what the function does and to what.

You might not need a noun for methods that are part of a class or module. A reset() method in a SatelliteConnection class or an open() function in the webbrowser module already provides the necessary context. You can tell that a satellite connection is the item being reset and that a web browser is the item being opened.

It's better to use long, descriptive names rather than an acronym or a name that's too short. A mathematician might immediately understand that a function named gcd() returns the greatest common denominator of two numbers, but everyone else would find getGreatestCommonDenominator() more informative.

Remember not to use any of Python's built-in function or module names, such as all, any, date, email, file, format, hash, id, input, list, min, max, object, open, random, set, str, sum, test, and type.

Function Size Trade-Offs

Some programmers say that functions should be as short as possible and no longer than what can fit on a single screen. A function that is only a dozen lines long is relatively easy to understand, at least compared to one that is hundreds of lines long. But making functions shorter by splitting up their code into multiple smaller functions can also have its downsides. Let's look at some of the advantages of small functions:

- The function's code is easier to understand.
- The function likely requires fewer parameters.
- The function is less likely to have side effects, as described in "Functional Programming" on page 172.
- The function is easier to test and debug.
- The function likely raises fewer different kinds of exceptions.

But there are also some disadvantages to short functions:

- Writing short functions often means a larger number of functions in the program.
- Having more functions means the program is more complicated.
- Having more functions also means having to come up with additional descriptive, accurate names, which is a difficult task.
- Using more functions requires you to write more documentation.
- The relationships between functions become more complicated.

Some people take the guideline "the shorter, the better" to an extreme and claim that all functions should be three or four lines of code at most. This is madness. For example, here's the getPlayerMove() function from Chapter 14's Tower of Hanoi game. The specifics of how this code works are unimportant. Just look at the function's general structure:

```
def getPlayerMove(towers):
    """Asks the player for a move. Returns (fromTower, toTower)."""

    while True:  # Keep asking player until they enter a valid move.
        print('Enter the letters of "from" and "to" towers, or QUIT.')
        print("(e.g. AB to moves a disk from tower A to tower B.)")
        print()
        response = input("> ").upper().strip()

        if response == "QUIT":
            print("Thanks for playing!")
            sys.exit()

        # Make sure the user entered valid tower letters:
        if response not in ("AB", "AC", "BA", "BC", "CA", "CB"):
            print("Enter one of AB, AC, BA, BC, CA, or CB.")
            continue  # Ask player again for their move.

        # Use more descriptive variable names:
        fromTower, toTower = response[0], response[1]

        if len(towers[fromTower]) == 0:
            # The "from" tower cannot be an empty tower:
            print("You selected a tower with no disks.")
            continue  # Ask player again for their move.
        elif len(towers[toTower]) == 0:
            # Any disk can be moved onto an empty "to" tower:
            return fromTower, toTower
        elif towers[toTower][-1] < towers[fromTower][-1]:
            print("Can't put larger disks on top of smaller ones.")
            continue  # Ask player again for their move.
        else:
            # This is a valid move, so return the selected towers:
            return fromTower, toTower
```

This function is 34 lines long. Although it covers multiple tasks, including allowing the player to enter a move, checking whether this move is valid, and asking the player again to enter a move if the move is invalid, these tasks all fall under the umbrella of getting the player's move. On the other hand, if we were devoted to writing short functions, we could break the code in getPlayerMove() into smaller functions, like this:

```
def getPlayerMove(towers):
    """Asks the player for a move. Returns (fromTower, toTower)."""

    while True:  # Keep asking player until they enter a valid move.
        response = askForPlayerMove()
```

```
            terminateIfResponseIsQuit(response)
            if not isValidTowerLetters(response):
                continue # Ask player again for their move.

            # Use more descriptive variable names:
            fromTower, toTower = response[0], response[1]

            if towerWithNoDisksSelected(towers, fromTower):
                continue   # Ask player again for their move.
            elif len(towers[toTower]) == 0:
                # Any disk can be moved onto an empty "to" tower:
                return fromTower, toTower
            elif largerDiskIsOnSmallerDisk(towers, fromTower, toTower):
                continue   # Ask player again for their move.
            else:
                # This is a valid move, so return the selected towers:
                return fromTower, toTower

def askForPlayerMove():
    """Prompt the player, and return which towers they select."""
    print('Enter the letters of "from" and "to" towers, or QUIT.')
    print("(e.g. AB to moves a disk from tower A to tower B.)")
    print()
    return input("> ").upper().strip()

def terminateIfResponseIsQuit(response):
    """Terminate the program if response is 'QUIT'"""
    if response == "QUIT":
        print("Thanks for playing!")
        sys.exit()

def isValidTowerLetters(towerLetters):
    """Return True if `towerLetters` is valid."""
    if towerLetters not in ("AB", "AC", "BA", "BC", "CA", "CB"):
        print("Enter one of AB, AC, BA, BC, CA, or CB.")
        return False
    return True

def towerWithNoDisksSelected(towers, selectedTower):
    """Return True if `selectedTower` has no disks."""
    if len(towers[selectedTower]) == 0:
        print("You selected a tower with no disks.")
        return True
    return False

def largerDiskIsOnSmallerDisk(towers, fromTower, toTower):
    """Return True if a larger disk would move on a smaller disk."""
    if towers[toTower][-1] < towers[fromTower][-1]:
        print("Can't put larger disks on top of smaller ones.")
        return True
    return False
```

These six functions are 56 lines long, nearly double the line count of the original code, but they do the same tasks. Although each function is easier to understand than the original getPlayerMove() function, the group of them together represents an increase in complexity. Readers of your code might have trouble understanding how they all fit together. The getPlayerMove() function is the only one called by other parts of the program; the other five functions are called only once, from getPlayerMove(). But the mass of functions doesn't convey this fact.

I also had to come up with new names and docstrings (the triple-quoted strings under each def statement, further explained in Chapter 11) for each new function. This leads to functions with confusingly similar names, such as getPlayerMove() and askForPlayerMove(). Also, getPlayerMove() is still longer than three or four lines, so if I were following the guideline "the shorter, the better," I'd need to split it into even smaller functions!

In this case, the policy of allowing only incredibly short functions might have resulted in simpler functions, but the overall complexity of the program increased drastically. In my opinion, functions should be fewer than 30 lines ideally and definitely no longer than 200 lines. Make your functions as short as reasonably possible but not any shorter.

Function Parameters and Arguments

A function's parameters are the variable names between the parentheses of the function's def statement, whereas the arguments are the values between a function call's parentheses. The more parameters a function has, the more configurable and generalized its code can be. But more parameters also mean greater complexity.

A good rule to adhere to is that zero to three parameters is fine, but more than five or six is probably too many. Once functions become overly complicated, it's best to consider how to split them into smaller functions with fewer parameters.

Default Arguments

One way to reduce the complexity of your function's parameters is by providing default arguments for your parameters. A *default argument* is a value used as an argument if the function call doesn't specify one. If the majority of function calls use a particular parameter value, we can make that value a default argument to avoid having to enter it repeatedly in the function call.

We specify a default argument in the def statement, following the parameter name and an equal sign. For example, in this introduction() function, a parameter named greeting has the value 'Hello' if the function call doesn't specify it:

```
>>> def introduction(name, greeting='Hello'):
...     print(greeting + ', ' + name)
...
>>> introduction('Alice')
```

```
Hello, Alice
>>> introduction('Hiro', 'Ohiyo gozaimasu')
Ohiyo gozaimasu, Hiro
```

When the introduction() function is called without a second argument, it uses the string 'Hello' by default. Note that parameters with default arguments must always come after parameters without default arguments.

Recall from Chapter 8 that you should avoid using a mutable object, such as an empty list [] or empty dictionary {}, as the default value. "Don't Use Mutable Values for Default Arguments" on page 143 explains the problem that this approach causes and its solution.

Using * and ** to Pass Arguments to Functions

You can use the * and ** syntax (often pronounced as *star* and *star star*) to pass groups of arguments to functions separately. The * syntax allows you to pass in the items in an iterable object (such as a list or tuple). The ** syntax allows you to pass in the key-value pairs in a mapping object (such as a dictionary) as individual arguments.

For example, the print() function can take multiple arguments. It places a space in between them by default, as the following code shows:

```
>>> print('cat', 'dog', 'moose')
cat dog moose
```

These arguments are called *positional arguments*, because their position in the function call determines which argument is assigned to which parameter. But if you stored these strings in a list and tried to pass the list, the print() function would think you were trying to print the list as a single value:

```
>>> args = ['cat', 'dog', 'moose']
>>> print(args)
['cat', 'dog', 'moose']
```

Passing the list to print() displays the list, including brackets, quotes, and comma characters.

One way to print the individual items in the list would be to split the list into multiple arguments by passing each item's index to the function individually, resulting in code that is harder to read:

```
>>> # An example of less readable code:
>>> args = ['cat', 'dog', 'moose']
>>> print(args[0], args[1], args[2])
cat dog moose
```

There's an easier way to pass these items to print(). You can use the * syntax to interpret the items in a list (or any other iterable data type) as individual positional arguments. Enter the following example into the interactive shell.

```
>>> args = ['cat', 'dog', 'moose']
>>> print(*args)
cat dog moose
```

The * syntax allows you pass the list items to a function individually, no matter how many items are in the list.

You can use the ** syntax to pass mapping data types (such as dictionaries) as individual *keyword arguments*. Keyword arguments are preceded by a parameter name and equal sign. For example, the print() function has a sep keyword argument that specifies a string to put in between the arguments it displays. It's set to a single space string ' ' by default. You can assign a keyword argument to a different value using either an assignment statement or the ** syntax. To see how this works, enter the following into the interactive shell:

```
>>> print('cat', 'dog', 'moose', sep='-')
cat-dog-moose
>>> kwargsForPrint = {'sep': '-'}
>>> print('cat', 'dog', 'moose', **kwargsForPrint)
cat-dog-moose
```

Notice that these instructions produce identical output. In the example, we used only one line of code to set up the kwargsForPrint dictionary. But for more complex cases, you might need more code to set up a dictionary of keyword arguments. The ** syntax allows you to create a custom dictionary of configuration settings to pass to a function call. This is useful especially for functions and methods that accept a large number of keyword arguments.

By modifying a list or dictionary at runtime, you can supply a variable number of arguments for a function call using the * and ** syntax.

Using * to Create Variadic Functions

You can also use the * syntax in def statements to create *variadic* or *varargs* functions that receive a varying number of positional arguments. For instance, print() is a variadic function, because you can pass any number of strings to it: print('Hello!') or print('My name is', name), for example. Note that although we used the * syntax in function calls in the previous section, we use the * syntax in function definitions in this section.

Let's look at an example by creating a product() function that takes any number of arguments and multiplies them together:

```
>>> def product(*args):
...     result = 1
...     for num in args:
...         result *= num
...     return result
...
>>> product(3, 3)
9
>>> product(2, 1, 2, 3)
12
```

Inside the function, args is just a regular Python tuple containing all the positional arguments. Technically, you can name this parameter anything, as long as it begins with the star (*), but it's usually named args by convention.

Knowing when to use the * takes some thought. After all, the alternative to making a variadic function is to have a single parameter that accepts a list (or other iterable data type), which contains a varying number of items. This is what the built-in sum() function does:

```
>>> sum([2, 1, 2, 3])
8
```

The sum() function expects one iterable argument, so passing it multiple arguments results in an exception:

```
>>> sum(2, 1, 2, 3)
Traceback (most recent call last):
  File "<stdin>", line 1, in <module>
TypeError: sum() takes at most 2 arguments (4 given)
```

Meanwhile, the built-in min() and max() functions, which find the minimum or maximum value of several values, accept a single iterable argument or multiple separate arguments:

```
>>> min([2, 1, 3, 5, 8])
1
>>> min(2, 1, 3, 5, 8)
1
>>> max([2, 1, 3, 5, 8])
8
>>> max(2, 1, 3, 5, 8)
8
```

All of these functions take a varying number of arguments, so why are their parameters designed differently? And when should we design functions to take a single iterable argument or multiple separate arguments using the * syntax?

How we design our parameters depends on how we predict a programmer will use our code. The print() function takes multiple arguments because programmers more often pass a series of strings, or variables that contain strings, to it, as in print('My name is', name). It isn't as common to collect these strings into a list over several steps and then pass the list to print(). Also, if you passed a list to print(), the function would print that list value in its entirety, so you can't use it to print the individual values in the list.

There's no reason to call sum() with separate arguments because Python already uses the + operator for that. Because you can write code like 2 + 4 + 8, you don't need to be able to write code like sum(2, 4, 8). It makes sense that you must pass the varying number of arguments only as a list to sum().

The min() and max() functions allow both styles. If the programmer passes one argument, the function assumes it's a list or tuple of values to inspect. If the programmer passes multiple arguments, it assumes these are the values to inspect. These two functions commonly handle lists of values while the program is running, as in the function call min(allExpenses). They also deal with separate arguments the programmer selects while writing the code, such as in max(0, someNumber). Therefore, the functions are designed to accept both kinds of arguments. The following myMinFunction(), which is my own implementation of the min() function, demonstrates this:

```
def myMinFunction(*args):
    if len(args) == 1:
    ❶ values = args[0]
    else:
    ❷ values = args

    if len(values) == 0:
    ❸ raise ValueError('myMinFunction() args is an empty sequence')

❹ for i, value in enumerate(values):
        if i == 0 or value < smallestValue:
            smallestValue = value
    return smallestValue
```

The myMinFunction() uses the * syntax to accept a varying number of arguments as a tuple. If this tuple contains only one value, we assume it's a sequence of values to inspect ❶. Otherwise, we assume that args is a tuple of values to inspect ❷. Either way, the values variable will contain a sequence of values for the rest of the code to inspect. Like the actual min() function, we raise ValueError if the caller didn't pass any arguments or passed an empty sequence ❸. The rest of the code loops through values and returns the smallest value found ❹. To keep this example simple, myMinFunction() accepts only sequences like lists or tuples rather than any iterable value.

You might wonder why we don't always write functions to accept both ways of passing a varying number of arguments. The answer is that it's best to keep your functions as simple as possible. Unless both ways of calling the function are common, choose one over the other. If a function usually deals with a data structure created while the program is running, it's better to have it accept a single parameter. If a function usually deals with arguments that the programmer specifies while writing the code, it's better to use the * syntax to accept a varying number of arguments.

Using ** to Create Variadic Functions

Variadic functions can use the ** syntax, too. Although the * syntax in def statements represents a varying number of positional arguments, the ** syntax represents a varying number of optional keyword arguments.

If you define a function that could take numerous optional keyword arguments without using the ** syntax, your def statement could become unwieldy. Consider a hypothetical formMolecule() function, which has parameters for all 118 known elements:

```
>>> def formMolecule(hydrogen, helium, lithium, beryllium, boron, --snip--
```

Passing 2 for the hydrogen parameter and 1 for the oxygen parameter to return 'water' would also be burdensome and unreadable, because you'd have to set all of the irrelevant elements to zero:

```
>>> formMolecule(2, 0, 0, 0, 0, 0, 0, 1, 0, 0, 0, 0, 0, 0, 0, 0, 0 --snip--
'water'
```

You could make the function more manageable by using named keyword parameters that each have a default argument, freeing you from having to pass that parameter an argument in a function call.

NOTE *Although the terms* argument *and* parameter *are well defined, programmers tend to use* keyword argument *and* keyword parameter *interchangeably.*

For example, this def statement has default arguments of 0 for each of the keyword parameters:

```
>>> def formMolecule(hydrogen=0, helium=0, lithium=0, beryllium=0, --snip--
```

This makes calling formMolecule() easier, because you only need to specify arguments for parameters that have a different value than the default argument. You can also specify the keyword arguments in any order:

```
>>> formMolecule(hydrogen=2, oxygen=1)
'water'
>>> formMolecule(oxygen=1, hydrogen=2)
'water'
>>> formMolecule(carbon=8, hydrogen=10, nitrogen=4, oxygen=2)
'caffeine'
```

But you still have an unwieldy def statement with 118 parameter names. And what if new elements were discovered? You'd have to update the function's def statement along with any documentation of the function's parameters.

Instead, you can collect all the parameters and their arguments as key-value pairs in a dictionary using the ** syntax for keyword arguments. Technically, you can name the ** parameter anything, but it's usually named kwargs by convention:

```
>>> def formMolecules(**kwargs):
...     if len(kwargs) == 2 and kwargs['hydrogen'] == 2 and
                              kwargs['oxygen'] == 1:
...         return 'water'
...     # (rest of code for the function goes here)
...
>>> formMolecules(hydrogen=2, oxygen=1)
'water'
```

The ** syntax indicates that the kwargs parameter can handle all keyword arguments passed in a function call. They'll be stored as key-value pairs in a dictionary assigned to the kwargs parameter. As new chemical elements are discovered, you'd need to update the function's code but not its def statement, because all keyword arguments are put into kwargs:

```
❶ >>> def formMolecules(**kwargs):
❷ ...     if len(kwargs) == 1 and kwargs.get('unobtanium') == 12:
  ...         return 'aether'
  ...     # (rest of code for the function goes here)
  ...
>>> formMolecules(unobtanium=12)
'aether'
```

As you can see, the def statement ❶ is the same as before, and only the function's code ❷ needed updating. When you use the ** syntax, the def statement and the function calls become much simpler to write and still produce readable code.

Using * and ** to Create Wrapper Functions

A common use case for the * and ** syntax in def statements is to create wrapper functions, which pass on arguments to another function and return that function's return value. You can use the * and ** syntax to forward any and all arguments to the wrapped function. For example, we can create a printLowercase() function that wraps the built-in print() function. It relies on print() to do the real work but converts the string arguments to lowercase first:

```
❶ >>> def printLower(*args, **kwargs):
❷ ...     args = list(args)
  ...     for i, value in enumerate(args):
  ...         args[i] = str(value).lower()
❸ ...     return print(*args, **kwargs)
  ...
>>> name = 'Albert'
>>> printLower('Hello,', name)
hello, albert
>>> printLower('DOG', 'CAT', 'MOOSE', sep=', ')
dog, cat, moose
```

The printLower() function ❶ uses the * syntax to accept a varying number of positional arguments in a tuple assigned to the args parameter, whereas the ** syntax assigns any keyword arguments to a dictionary in the kwargs parameter. If a function uses *args and **kwargs together, the *args parameter must come before the **kwargs parameter. We pass these on to the wrapped print() function, but first our function modifies some of the arguments, so we create a list form of the args tuple ❷.

After changing the strings in args to lowercase, we pass the items in args and key-value pairs in kwargs as separate arguments to print() using the * and ** syntax ❸. The return value of print() also gets returned as the return value of printLower(). These steps effectively wrap the print() function.

Functional Programming

Functional programming is a programming paradigm that emphasizes writing functions that perform calculations without modifying global variables or any external state (such as files on the hard drive, internet connections, or databases). Some programming languages, such as Erlang, Lisp, and Haskell, are heavily designed around functional programming concepts. Although not shackled to the paradigm, Python has some functional programming features. The main ones that Python programs can use are side-effect-free functions, higher-order functions, and lambda functions.

Side Effects

Side effects are any changes a function makes to the parts of the program that exist outside of its own code and local variables. To illustrate this, let's create a subtract() function that implements Python's subtraction operator (-):

```
>>> def subtract(number1, number2):
...     return number1 - number2
...
>>> subtract(123, 987)
-864
```

This subtract() function has no side effects. That is, it doesn't affect anything in the program that isn't a part of its code. There's no way to tell from the program's or the computer's state whether the subtract() function has been called once, twice, or a million times before. A function might modify local variables inside the function, but these changes remain isolated from the rest of the program.

Now consider an addToTotal() function, which adds the numeric argument to a global variable named TOTAL:

```
>>> TOTAL = 0
>>> def addToTotal(amount):
...     global TOTAL
...     TOTAL += amount
...     return TOTAL
...
>>> addToTotal(10)
10
>>> addToTotal(10)
20
>>> addToTotal(9999)
10019
>>> TOTAL
10019
```

The addToTotal() function does have a side effect, because it modifies an element that exists outside of the function: the TOTAL global variable. Side effects can be more than changes to global variables. They include updating or deleting files, printing text onscreen, opening a database connection, authenticating to a server, or making any other change outside of the function. Any trace that a function call leaves behind after returning is a side effect.

Side effects can also include making in-place changes to mutable objects referred to outside of the function. For example, the following removeLastCatFromList() function modifies the list argument in-place:

```
>>> def removeLastCatFromList(petSpecies):
...     if len(petSpecies) > 0 and petSpecies[-1] == 'cat':
...         petSpecies.pop()
...
>>> myPets = ['dog', 'cat', 'bird', 'cat']
>>> removeLastCatFromList(myPets)
>>> myPets
['dog', 'cat', 'bird']
```

In this example, the myPets variable and petSpecies parameter hold references to the same list. Any in-place modifications made to the list object inside the function would also exist outside the function, making this modification a side effect.

A related concept, a *deterministic function*, always returns the same return value given the same arguments. The subtract(123, 987) function call always returns –864. Python's built-in round() function always returns 3 when passed 3.14 as an argument. A *nondeterministic function* won't always return the same values when passed the same arguments. For example, calling random.randint(1, 10) returns a random integer between 1 and 10. The time.time() function has no arguments, but it returns a different value depending on what your computer's clock is set to when the function was called. In the case of time.time(), the clock is an external resource that is effectively an input into the function the same way an argument is. Functions that depend on resources external to the function (including global variables, files on the hard drive, databases, and internet connections) are not considered deterministic.

One benefit of deterministic functions is that you can cache their values. There's no need for subtract() to calculate the difference of 123 and 987 more than once if it can remember the return value from the first time it's called with those arguments. Therefore, deterministic functions allow us to make a *space-time trade-off*, quickening the runtime of a function by using space in memory to cache previous results.

A function that is deterministic and free of side effects is called a *pure function*. Functional programmers strive to create only pure functions in their programs. In addition to those already noted, pure functions offer several benefits:

- They're well suited for unit testing, because they don't require you to set up any external resources.

- It's easy to reproduce bugs in a pure function by calling the function with the same arguments.

- Pure functions can call other pure functions and remain pure.

- In multithreaded programs, pure functions are thread-safe and can safely run concurrently. (Multithreading is beyond the scope of this book.)

- Multiple calls to pure functions can run on parallel CPU cores or in a multithreaded program because they don't have to depend on any external resources that require them to be run in any particular sequence.

You can and should write pure functions in Python whenever possible. Python functions are made pure by convention only; there's no setting that causes the Python interpreter to enforce purity. The most common way to make your functions pure is to avoid using global variables in them and ensure they don't interact with files, the internet, the system clock, random numbers, or other external resources.

Higher-Order Functions

Higher-order functions can accept other functions as arguments or return functions as return values. For example, let's define a function named callItTwice() that will call a given function twice:

```
>>> def callItTwice(func, *args, **kwargs):
...     func(*args, **kwargs)
...     func(*args, **kwargs)
...
>>> callItTwice(print, 'Hello, world!')
Hello, world!
Hello, world!
```

The callItTwice() function works with any function it's passed. In Python, functions are *first-class objects*, meaning they're like any other object: you can store functions in variables, pass them as arguments, or use them as return values.

Lambda Functions

Lambda functions, also known as *anonymous functions* or *nameless functions*, are simplified functions that have no names and whose code consists solely of one return statement. We often use lambda functions when passing functions as arguments to other functions.

For example, we could create a normal function that accepts a list containing a 4 by 10 rectangle's width and height, like this:

```
>>> def rectanglePerimeter(rect):
...     return (rect[0] * 2) + (rect[1] * 2)
...
>>> myRectangle = [4, 10]
>>> rectanglePerimeter(myRectangle)
28
```

The equivalent lambda function would look like this:

```
lambda rect: (rect[0] * 2) + (rect[1] * 2)
```

To define a Python lambda function, use the `lambda` keyword, followed by a comma-delimited list of parameters (if any), a colon, and then an expression that acts as the return value. Because functions are first-class objects, you can assign a lambda function to a variable, effectively replicating what a `def` statement does:

```
>>> rectanglePerimeter = lambda rect: (rect[0] * 2) + (rect[1] * 2)
>>> rectanglePerimeter([4, 10])
28
```

We assigned this lambda function to a variable named `rectanglePerimeter`, essentially giving us a `rectanglePerimeter()` function. As you can see, functions created by `lambda` statements are the same as functions created by `def` statements.

NOTE   *In real-world code, use `def` statements rather than assigning lambda functions to a constant variable. Lambda functions are specifically made for situations in which a function doesn't need a name.*

The lambda function syntax is helpful for specifying small functions to serve as arguments to other function calls. For example, the `sorted()` function has a keyword argument named key that lets you specify a function. Instead of sorting items in a list based on the item's value, it sorts them based on the function's return value. In the following example, we pass `sorted()` a lambda function that returns the perimeter of the given rectangle. This makes the `sorted()` function sort based on the calculated perimeter of its [width, height] list rather than based directly on the [width, height] list:

```
>>> rects = [[10, 2], [3, 6], [2, 4], [3, 9], [10, 7], [9, 9]]
>>> sorted(rects, key=lambda rect: (rect[0] * 2) + (rect[1] * 2))
[[2, 4], [3, 6], [10, 2], [3, 9], [10, 7], [9, 9]]
```

Rather than sorting the values [10, 2] or [3, 6], for example, the function now sorts based on the returned perimeter integers 24 and 18. Lambda functions are a convenient syntactic shortcut: you can specify a small one-line lambda function instead of defining a new, named function with a `def` statement.

Mapping and Filtering with List Comprehensions

In earlier Python versions, the `map()` and `filter()` functions were common higher-order functions that could transform and filter lists, often with the help of lambda functions. Mapping could create a list of values based on the values of another list. Filtering could create a list that contained only the values from another list that match some criteria.

For example, if you wanted to create a new list that had strings instead of the integers [8, 16, 18, 19, 12, 1, 6, 7], you could pass that list and lambda n: str(n) to the map() function:

```
>>> mapObj = map(lambda n: str(n), [8, 16, 18, 19, 12, 1, 6, 7])
>>> list(mapObj)
['8', '16', '18', '19', '12', '1', '6', '7']
```

The map() function returns a map object, which we can get in list form by passing it to the list() function. The mapped list now contains string values based on the original list's integer values. The filter() function is similar, but here, the lambda function argument determines which items in the list remain (if the lambda function returns True) or are filtered out (if it returns False). For example, we could pass lambda n: n % 2 == 0 to filter out any odd integers:

```
>>> filterObj = filter(lambda n: n % 2 == 0, [8, 16, 18, 19, 12, 1, 6, 7])
>>> list(filterObj)
[8, 16, 18, 12, 6]
```

The filter() function returns a filter object, which we can once again pass to the list() function. Only the even integers remain in the filtered list.

But the map() and filter() functions are outdated ways to create mapped or filtered lists in Python. Instead, you can now create them with list comprehensions. List comprehensions not only free you from writing out a lambda function, but are also faster than map() and filter().

Here we replicate the map() function example using a list comprehension:

```
>>> [str(n) for n in [8, 16, 18, 19, 12, 1, 6, 7]]
['8', '16', '18', '19', '12', '1', '6', '7']
```

Notice that the str(n) part of the list comprehension is similar to lambda n: str(n).

And here we replicate the filter() function example using a list comprehension:

```
>>> [n for n in [8, 16, 18, 19, 12, 1, 6, 7] if n % 2 == 0]
[8, 16, 18, 12, 6]
```

Notice that the if n % 2 == 0 part of the list comprehension is similar to lambda n: n % 2 == 0.

Many languages have a concept of functions as first-class objects, allowing for the existence of higher-order functions, including mapping and filtering functions.

Return Values Should Always Have the Same Data Type

Python is a dynamically typed language, which means that Python functions and methods are free to return values of any data type. But to make your functions more predictable, you should strive to have them return values of only a single data type.

For example, here's a function that, depending on a random number, returns either an integer value or a string value:

```
>>> import random
>>> def returnsTwoTypes():
...     if random.randint(1, 2) == 1:
...         return 42
...     else:
...         return 'forty two'
```

When you're writing code that calls this function, it can be easy to forget that you must handle several possible data types. To continue this example, say we call returnsTwoTypes() and want to convert the *number* that it returns to hexadecimal:

```
>>> hexNum = hex(returnsTwoTypes())
>>> hexNum
'0x2a'
```

Python's built-in hex() function returns a string of a hexadecimal number of the integer value it was passed. This code works fine as long as returnsTwoTypes() returns an integer, giving us the impression that this code is bug free. But when returnsTwoTypes() returns a string, it raises an exception:

```
>>> hexNum = hex(returnsTwoTypes())
Traceback (most recent call last):
  File "<stdin>", line 1, in <module>
TypeError: 'str' object cannot be interpreted as an integer
```

Of course, we should always remember to handle every possible data type that the return value could have. But in the real world, it's easy to forget this. To prevent these bugs, we should always attempt to make functions return values of a single data type. This isn't a strict requirement, and sometimes there's no way around having your function return values of different data types. But the closer you get to returning only one type, the simpler and less bug prone your functions will be.

There is one case in particular to be aware of: don't return None from your function unless your function always returns None. The None value is the only value in the NoneType data type. It's tempting to have a function return None to signify that an error occurred (I discuss this practice in the next section, "Raising Exceptions vs. Returning Error Codes"), but you should reserve returning None for functions that have no meaningful return value.

The reason is that returning None to indicate an error is a common source of uncaught 'NoneType' object has no attribute exceptions:

```
>>> import random
>>> def sometimesReturnsNone():
...     if random.randint(1, 2) == 1:
...         return 'Hello!'
...     else:
...         return None
...
>>> returnVal = sometimesReturnsNone()
>>> returnVal.upper()
'HELLO!'
>>> returnVal = sometimesReturnsNone()
>>> returnVal.upper()
Traceback (most recent call last):
  File "<stdin>", line 1, in <module>
AttributeError: 'NoneType' object has no attribute 'upper'
```

This error message is rather vague, and it could take some effort to trace its cause back to a function that normally returns an expected result but could also return None when an error happens. The problem occurred because sometimesReturnsNone() returned None, which we then assigned to the returnVal variable. But the error message would lead you to think the problem occurred in the call to the upper() method.

In a 2009 conference talk, computer scientist Tony Hoare apologized for inventing the null reference (the general analogous value to Python's None value) in 1965, saying "I call it my billion dollar mistake. [...] I couldn't resist the temptation to put in a null reference, simply because it was so easy to implement. This has led to innumerable errors, vulnerabilities, and system crashes, which have probably caused a billion dollars of pain and damage in the last 40 years." You can view his full talk online at *https://autbor.com/ billiondollarmistake*.

Raising Exceptions vs. Returning Error Codes

In Python, the meanings of the terms *exception* and *error* are roughly the same: an exceptional circumstance in your program that usually indicates a problem. Exceptions became popular as a programming language feature in the 1980s and 1990s with C++ and Java. They replaced the use of *error codes*, which are values returned from functions to indicate a problem. The benefit of exceptions is that return values are only related to the function's purpose instead of also indicating the presence of errors.

Error codes can also cause issues in your program. For example, Python's find() string method normally returns the index where it found a substring, and if it's unable to find it, it returns -1 as an error code. But because we can also use -1 to specify the index from the end of a string, inadvertently using -1 as an error code might introduce a bug. Enter the following in the interactive shell to see how this works.

```
>>> print('Letters after b in "Albert":', 'Albert'['Albert'.find('b') + 1:])
Letters after b in "Albert": ert
>>> print('Letters after x in "Albert":', 'Albert'['Albert'.find('x') + 1:])
Letters after x in "Albert": Albert
```

The 'Albert'.find('x') part of the code evaluates to the error code -1. That makes the expression 'Albert'['Albert'.find('x') + 1:] evaluate to 'Albert'[-1 + 1:], which further evaluates to 'Albert'[0:] and then to 'Albert'. Obviously, this isn't the code's intended behavior. Calling index() instead of find(), as in 'Albert'['Albert'.index('x') + 1:], would have raised an exception, making the problem obvious and unignorable.

The index() string method, on the other hand, raises a ValueError exception if it's unable to find a substring. If you don't handle this exception, it will crash the program—behavior that is often preferable to not noticing the error.

The names of exception classes often end with "Error" when the exception indicates an actual error, such as ValueError, NameError, or SyntaxError. Exception classes that represent exceptional cases that aren't necessarily errors include StopIteration, KeyboardInterrupt, or SystemExit.

Summary

Functions are a common way of grouping our programs' code together, and they require you to make certain decisions: what to name them, how big to make them, how many parameters they should have, and how many arguments you should pass for those parameters. The * and ** syntax in def statements allows functions to receive a varying number of parameters, making them variadic functions.

Although not a functional programming language, Python has many features that functional programming languages use. Functions are first-class objects, meaning you can store them in variables and pass them as arguments to other functions (which are called higher-order functions in this context). Lambda functions offer a short syntax for specifying nameless, anonymous functions as the arguments for higher-order functions. The most common higher-order functions in Python are map() and filter(), although you can execute the functionality they provide faster with list comprehensions.

The return values of your functions should always be the same data type. You shouldn't use return values as error codes: exceptions are for indicating errors. The None value in particular is often mistakenly used as an error code.

11

COMMENTS, DOCSTRINGS, AND TYPE HINTS

The comments and documentation in your source code can be just as important as the code. The reason is that software is never finished; you'll always need to make changes, whether you're adding new features or fixing bugs. But you can't change code unless you understand it, so it's important that you keep it in a readable state. As computer scientists Harold Abelson, Gerald Jay Sussman, and Julie Sussman once wrote, "Programs must be written for people to read, and only incidentally for machines to execute."

Comments, docstrings, and type hints help you maintain your code's legibility. Comments are short, plain-English explanations that you write directly in the source code and the computer ignores them. Comments offer helpful notes, warnings, and reminders to others who didn't write the code, or sometimes even to the code's programmer in the future. Almost every programmer has asked themselves, "Who wrote this unreadable mess?" only to find the answer is, "I did."

Docstrings are a Python-specific form of documentation for your functions, methods, and modules. When you specify comments in the docstring format, automated tools, such as documentation generators or Python's built-in help() module, make it easy for developers to find information about your code.

Type hints are directives you can add to your Python source code to specify the data types of variables, parameters, and return values. This allows static code analysis tools to verify that your code won't generate any exceptions due to incorrectly typed values. Type hints first appeared in Python 3.5, but because they're based on comments, you can use them in any Python version.

This chapter focuses on the aforementioned three techniques for embedding documentation inside your code to make it more readable. External documentation, such as user manuals, online tutorials, and reference materials, are important but not covered in this book. If you want to learn more about external documentation, check out the Sphinx documentation generator at *https://www.sphinx-doc.org/*.

Comments

Like most programming languages, Python supports single-line comments and multiline comments. Any text that appears between the number sign # and the end of the line is a single-line comment. Although Python doesn't have a dedicated syntax for multiline comments, a triple-quotes multiline string can serve as one. After all, a string value by itself doesn't cause the Python interpreter to do anything. Look at this example:

```
# This is a single-line comment.

"""This is a
multiline string that
also works as a multiline comment. """
```

If your comment spans multiple lines, it's better to use a single multiline comment than several consecutive single-line comments, which are harder to read, as you can see here:

```
"""This is a good way
to write a comment
that spans multiple lines. """
# This is not a good way
# to write a comment
# that spans multiple lines.
```

Comments and documentation are often afterthoughts in the programming process or are even considered by some to do more harm than good. But as "Myth: Comments Are Unnecessary" on page 83 explained, comments are not optional if you want to write professional, readable code. In this section, we'll write useful comments that inform the reader without detracting from the program's readability.

Comment Style

Let's look at some comments that follow good style practices:

```
❶ # Here is a comment about this code:
someCode()

❷ # Here is a lengthier block comment that spans multiple lines using
# several single-line comments in a row.
❸ #
# These are known as block comments.

if someCondition:
❹     # Here is a comment about some other code:
❺     someOtherCode()  # Here is an inline comment.
```

Comments should generally exist on their own line rather than at the end of a line of code. Most of the time, they should be complete sentences with appropriate capitalization and punctuation rather than phrases or single words ❶. The exception is that comments should obey the same line-length limits that the source code does. Comments that span multiple lines ❷ can use multiple single-line comments in a row, known as *block comments*. We separate paragraphs in block comments using a blank, single-line comment ❸. Comments should have the same level of indentation as the code they're commenting ❹. Comments that follow a line of code are called *inline comments* ❺ and at least two spaces should separate the code from the comment.

Single-line comments should have one space after the # sign:

```
#Don't write comments immediately after the # sign.
```

Comments can include links to URLs with related information, but links should never replace comments because linked content could disappear from the internet at any time:

```
# Here is a detailed explanation about some aspect of the code
# that is supplemented by a URL. More info at https://example.com
```

The aforementioned conventions are matters of style rather than content, but they contribute to the comments' readability. The more readable your comments are, the more likely programmers will pay attention to them, and comments are only useful if programmers read them.

Inline Comments

Inline comments come at the end of a line of code, as in the following case:

```
while True:  # Keep asking player until they enter a valid move.
```

Inline comments are brief so they fit within the line-length limits set by the program's style guide. This means they can easily end up being too short to provide enough information. If you do decide to use inline comments, make sure the comment describes only the line of code it immediately follows. If your inline comment requires more space or describes additional lines of code, put it on its own line.

One common and appropriate use of inline comments is to explain the purpose of a variable or give some other kind of context for it. These inline comments are written on the assignment statement that creates the variable:

```
TOTAL_DISKS = 5  # More disks means a more difficult puzzle.
```

Another common use of inline comments is to add context about the values of variables when you create them:

```
month = 2  # Months range from 0 (Jan) to 11 (Dec).
catWeight = 4.9  # Weight is in kilograms.
website = 'inventwithpython.com'  # Don't include "https://" at front.
```

Inline comments should not specify the variable's data type, because this is obvious from the assignment statement, unless it's done in the comment form of a type hint, as described in "Backporting Type Hints with Comments" later in this chapter.

Explanatory Comments

In general, comments should explain why code is written the way it is rather than what the code does or how it does what it does. Even with the proper code style and useful naming conventions covered in Chapters 3 and 4, the actual code can't explain the original programmer's intentions. If you wrote the code, you might even forget details about it after a few weeks. Present You should write informative code comments to prevent Future You from cursing Past You.

For example, here's an unhelpful comment that explains what the code is doing. Rather than hinting at the motivation for this code, it states the obvious:

```
>>> currentWeekWages *= 1.5  # Multiply the current week's wages by 1.5
```

This comment is worse than useless. It's obvious from the code that the currentWeekWages variable is being multiplied by 1.5, so omitting the comment entirely would simplify your code. The following would be a much better comment:

```
>>> currentWeekWages *= 1.5  # Account for time-and-a-half wage rate.
```

This comment explains the intention behind this line of code rather than repeating how the code works. It provides context that even well-written code cannot.

Summary Comments

Explaining the programmer's intent isn't the only way comments can be useful. Brief comments that summarize several lines of code allow the reader to skim the source code and get a general idea of what it does. Programmers often use a blank space to separate "paragraphs" of code from each other, and the summarizing comments usually occupy one line at the start of these paragraphs. Unlike one-line comments that explain single lines of code, the summarizing comments describe what the code does at a higher level of abstraction.

For example, you can tell from reading these four lines of code that they set the `playerTurn` variable to a value representing the opposite player. But the short, single-line comment spares the reader from having to read and reason about the code to understand the purpose of doing this:

```
# Switch turns to other player:
if playerTurn == PLAYER_X:
    playerTurn = PLAYER_O
elif playerTurn == PLAYER_O:
    playerTurn = PLAYER_X
```

A scattering of these summary comments throughout your program makes it much easier to skim. The programmer can then take a closer look at any particular points of interest. Summary comments can also prevent programmers from developing a misleading idea about what the code does. A brief, summarizing comment can confirm that the developer properly understood how the code works.

"Lessons Learned" Comments

When I worked at a software company, I was once asked to adapt a graph library so it could handle the real-time updates of millions of data points in a chart. The library we were using could either update graphs in real time *or* support graphs with millions of data points, but not both. I thought I could finish the task in a few days. By the third week, I was still convinced that I could finish in a few days. Each day, the solution seemed to be just around the corner, and during the fifth week I had a working prototype.

During this entire process, I learned a lot of details about how the graphing library worked and what its capabilities and limitations were. I then spent a few hours writing up these details into a page-long comment that I placed in the source code. I knew that anyone else who needed to make changes to my code later would encounter all the same, seemingly simple issues I had, and that the documentation I was writing would save them literally weeks of effort.

These *lessons-learned* comments, as I call them, might span several paragraphs, making them seem out of place in a source code file. But the information they contain is gold for anyone who needs to maintain this code. Don't be afraid to write lengthy, detailed comments in your source code file to explain how something works. To other programmers, many of these details will be unknown, misunderstood, or easily overlooked. Software developers who don't need them can easily skip past them, but developers who do need them will be grateful for them. Remember that, as with other comments, a lessons-learned comment is not the same as module or function documentation (which docstrings handle). It also isn't a tutorial or how-to guide aimed at users of the software. Instead, lessons-learned comments are for developers reading the source code.

Because my lessons-learned comment concerned an open source graph library and could be useful to others, I took a moment to post it as an answer to the public question-and-answer site *https://stackoverflow.org*, where others in a similar situation could find it.

Legal Comments

Some software companies or open source projects have a policy of including copyright, software license, and authorship information in comments at the top of each source code file for legal reasons. These annotations should consist of a few lines at most and look something like this:

```
"""Cat Herder 3.0 Copyright (C) 2021 Al Sweigart. All rights reserved.
See license.txt for the full text."""
```

If possible, refer to an external document or website that contains the full text of the license rather than including the entire lengthy license at the top of every source code file. It's tiring to have to scroll past several screen lengths of text whenever you open a source code file, and including the full license doesn't provide additional legal protection.

Professional Comments

At my first software job, a senior co-worker I greatly respected took me aside and explained that because we sometimes released our products' source code to clients, it was important that the comments maintain a professional tone. Apparently, I had written "WTF" in one of the comments for an especially frustrating part of the code. I felt embarrassed, immediately apologized, and edited the comment. Since that moment, I've kept my code, even for personal projects, at a certain level of professionalism.

You might be tempted to add levity or vent your frustrations in your program's comments, but make it a habit to avoid doing so. You don't know who will read your code in the future, and it's easy to misinterpret the tone of a text. As explained in "Avoid Jokes, Puns, and Cultural References" on page 64, the best policy is to write your comments in a polite, direct, and humorless tone.

Codetags and TODO Comments

Programmers sometimes leave short comments to remind themselves about work that remains to be done. This usually takes the form of a *codetag*: a comment with an all-uppercase label, such as TODO, followed by a short description. Ideally, you would use project management tools to track these sorts of issues rather than burying them deep in the source code. But for smaller, personal projects that aren't using such tools, the occasional TODO comment can serve as a helpful reminder. Here's an example:

```
_chargeIonFluxStream()  # TODO: Investigate why this fails every Tuesday.
```

You can use a few different codetags for these reminders:

TODO   Introduces a general reminder about work that needs to be done

FIXME   Introduces a reminder that this part of the code doesn't entirely work

HACK   Introduces a reminder that this part of the code works, perhaps barely, but that the code should be improved

XXX   Introduces a general warning, often of high severity

You should follow these always-uppercase labels with more specific descriptions of the task or problem at hand. Later, you can search the source code for the labels to find the code that needs fixing. The downside is that you can easily forget about these reminders unless you happen to be reading the section of your code that they're in. Codetags shouldn't replace a formal issue tracker or bug reporter tool. If you do use a codetag in your code, I recommend keeping it simple: use only TODO and forgo the others.

Magic Comments and Source File Encoding

You might have seen *.py* source files with something like the following lines at the top:

```
❶ #!/usr/bin/env python3
❷ # -*- coding: utf-8 -*-
```

These *magic comments*, which always appear at the top of the file, provide interpreter or encoding information. The shebang line ❶ (introduced in Chapter 2) tells your operating system which interpreter to use to run the instructions in the file.

The second magic comment is an *encoding definition* line ❷. In this case, the line defines UTF-8 as the Unicode encoding scheme to use for the source file. You almost never need to include this line, because most editors and IDEs already save source code files in the UTF-8 encoding, and Python versions starting with Python 3.0 treat UTF-8 as the defined encoding by default. Files encoded in UTF-8 can contain any character, so your *.py* source file will be just fine whether it includes English, Chinese, or Arabic letters.

For an introduction to Unicode and string encodings, I highly recommend Ned Batchelder's blog post, "Pragmatic Unicode" at *https://nedbatchelder.com/text/unipain.html*.

Docstrings

Docstrings are multiline comments that appear either at the top of a module's *.py* source code file or directly following a class or def statement. They provide documentation about the module, class, function, or method being defined. Automated documentation generator tools use these docstrings to generate external documentation files, such as help files or web pages.

Docstrings must use triple-quoted, multiline comments rather than single-line comments that begin with a hash mark, #. Docstrings should always use three double quotes for its triple-quoted strings rather than three single quotes. For example, here is an excerpt from the *sessions.py* file in the popular requests module:

```
❶ # -*- coding: utf-8 -*-

❷ """
requests.session
~~~~~~~~~~~~~~~~~

This module provides a Session object to manage and persist settings across
requests (cookies, auth, proxies).
"""
import os
import sys
--snip-
class Session(SessionRedirectMixin):
    ❸ """A Requests session.

    Provides cookie persistence, connection-pooling, and configuration.

    Basic Usage::

      >>> import requests
      >>> s = requests.Session()
      >>> s.get('https://httpbin.org/get')
      <Response [200]>
--snip--

    def get(self, url, **kwargs):
        ❹ r"""Sends a GET request. Returns :class:`Response` object.

        :param url: URL for the new :class:`Request` object.
        :param \*\*kwargs: Optional arguments that ``request`` takes.
        :rtype: requests.Response
        """

--snip--
```

The *sessions.py* file's request contains docstrings for the module ❷, the Session class ❸, and the Session class's get() method ❹. Note that although the module's docstring must be the first string to appear in the module, it should come after any magic comments, such as the shebang line or encoding definition ❶.

Later, you can retrieve the docstring for a module, class, function, or method by checking the respective object's __doc__ attribute. For example, here we examine the docstrings to find out more about the sessions module, the Session class, and the get() method:

```
>>> from requests import sessions
>>> sessions.__doc__
'\nrequests.session\n~~~~~~~~~~~~~~~~~~\n\nThis module provides a Session object
to manage and persist settings across\nrequests (cookies, auth, proxies).\n'
>>> sessions.Session.__doc__
"A Requests session.\n\n    Provides cookie persistence, connection-pooling,
and configuration.\n\n    Basic Usage::\n\n        >>> import requests\n
--snip--
>>> sessions.Session.get.__doc__
'Sends a GET request. Returns :class:`Response` object.\n\n        :param url:
URL for the new :class:`Request` object.\n        :param \\*\\*kwargs:
--snip--
```

Automated documentation tools can take advantage of docstrings to provide context-appropriate information. One of these tools is Python's built-in help() function, which displays the docstring of the object you pass it in a more readable format than the raw __doc__ string directly. This is useful when you're experimenting in the interactive shell, because you can immediately pull up information on any modules, classes, or functions you're trying to use:

```
>>> from requests import sessions
>>> help(sessions)
Help on module requests.sessions in requests:

NAME
    requests.sessions

DESCRIPTION
    requests.session
    ~~~~~~~~~~~~~~~~~

    This module provides a Session object to manage and persist settings
-- More  --
```

If the docstring is too large to fit onscreen, Python displays -- More -- at the bottom of the window. You can press ENTER to scroll to the next line, press the spacebar to scroll to the next page, or press Q to quit viewing the docstring.

Generally speaking, a docstring should contain a single line that summarizes the module, class, or function, followed by a blank line and more detailed information. For functions and methods, this can include information about their parameters, return value, and side effects. We write docstrings for other programmers rather than users of the software, so they should contain technical information, not tutorials.

Docstrings provide a second key benefit because they integrate documentation into the source code. When you write documentation separate from code, you can often forget about it entirely. Instead, when you place docstrings at the top of the modules, classes, and functions, the information remains easy to review and update.

You might not always immediately be able to write docstrings if you're still working on the code it's meant to describe. In that case, include a TODO comment in the docstring as a reminder to fill in the rest of the details. For example, the following fictional reverseCatPolarity() function has a poor docstring that states the obvious:

```
def reverseCatPolarity(catId, catQuantumPhase, catVoltage):
    """Reverses the polarity of a cat.

    TODO Finish this docstring."""
--snip--
```

Because every class, function, and method should have a docstring, you might be tempted to write only the minimal amount of documentation and move on. Without a TODO comment, it's easy to forget that this docstring will eventually need rewriting.

PEP 257 contains further documentation on docstrings at *https://www .python.org/dev/peps/pep-0257/*.

Type Hints

Many programming languages have *static typing*, meaning that programmers must declare the data types of all variables, parameters, and return values in the source code. This allows the interpreter or compiler to check that the code uses all objects correctly before the program runs. Python has *dynamic typing*: variables, parameters, and return values can be of any data type or even change data types while the program runs. Dynamic languages are often easier to program with because they require less formal specification, but they lack the bug-preventing advantages that static languages have. If you write a line of Python code, such as round('forty two'), you might not realize that you're passing a string to a function that accepts only int or float arguments until you run the code and it causes an error. A statically typed language gives you an early warning when you assign a value or pass an argument of the wrong type.

Python's *type hints* offer optional static typing. In the following example, the type hints are in bold:

```
def describeNumber(number: int) -> str:
    if number % 2 == 1:
        return 'An odd number. '
```

```
    elif number == 42:
        return 'The answer. '
    else:
        return 'Yes, that is a number. '

myLuckyNumber: int = 42
print(describeNumber(myLuckyNumber))
```

As you can see, for parameters or variables, the type hint uses a colon to separate the name from the type, whereas for return values, the type hint uses an arrow (->) to separate the def statement's closing parentheses from the type. The describeNumber() function's type hints show that it takes an integer value for its number parameter and returns a string value.

If you use type hints, you don't have to apply them to every bit of data in your program. Instead, you could use a *gradual typing* approach, which is a compromise between the flexibility of dynamic typing and the safety of static typing in which you include type hints for only certain variables, parameters, and return values. But the more type hinted your program is, the more information the static code analysis tool has to spot potential bugs in your program.

Notice in the preceding example that the names of the specified types match the names of the int() and str() constructor functions. In Python, *class*, *type*, and *data type* have the same meaning. For any instances made from classes, you should use the class name as the type:

```
import datetime
❶ noon: datetime.time = datetime.time(12, 0, 0)

class CatTail:
    def __init__(self, length: int, color: str) -> None:
        self.length = length
        self.color = color

❷ zophieTail: CatTail = CatTail(29, 'grey')
```

The noon variable has the type hint datetime.time ❶ because it's a time object (which is defined in the datetime module). Likewise, the zophieTail object has the CatTail type hint ❷ because it's an object of the CatTail class we created with a class statement. Type hints automatically apply to all sub-classes of the specified type. For example, a variable with the type hint dict could be set to any dictionary value but also to any collections.OrderedDict and collections.defaultdict values, because these classes are subclasses of dict. Chapter 16 covers subclasses in more detail.

Static type-checking tools don't necessarily need type hints for variables. The reason is that static type-checking tools do *type inference*, infer-ring the type from the variable's first assignment statement. For example, from the line spam = 42, the type checker can infer that spam is supposed to have a type hint of int. But I recommend setting a type hint anyway. A future change to a float, as in spam = 42.0, would also change the inferred type, which might not be your intention. It's better to force the programmer to change the type hint when changing the value to confirm that they've made an intentional rather than incidental change.

Using Static Analyzers

Although Python supports syntax for type hints, the Python interpreter completely ignores them. If you run a Python program that passes an invalidly typed variable to a function, Python will behave as though the type hints don't exist. In other words, type hints don't cause the Python interpreter to do any runtime type checking. They exist only for the benefit of static type-checking tools, which analyze the code before the program runs, not while the program is running.

We call these tools *static analysis* tools because they analyze the source code before the program runs, whereas runtime analysis or dynamic analysis tools analyze running programs. (Confusingly, *static* and *dynamic* in this case refer to whether the program is running, but *static typing* and *dynamic typing* refer to how we declare the data types of variables and functions. Python is a dynamically typed language that has static analysis tools, such as Mypy, written for it.)

Installing and Running Mypy

Although Python doesn't have an official type-checker tool, Mypy is currently the most popular third-party type checker. You can install Mypy with pip by running this command:

```
python -m pip install -user mypy
```

Run python3 instead of python on macOS and Linux. Other well-known type checkers include Microsoft's Pyright, Facebook's Pyre, and Google's Pytype.

To run the type checker, open a Command Prompt or Terminal window and run the python -m mypy command (to run the module as an application), passing it the filename of the Python code to check. In this example, I'm checking the code for an example program I created in a file named *example.py*:

```
C:\Users\Al\Desktop>python -m mypy example.py
Incompatible types in assignment (expression has type "float", variable has
type "int")
Found 1 error in 1 file (checked 1 source file)
```

The type checker outputs nothing if there are no problems and prints error messages if there are. In this *example.py* file, there's a problem on line 171, because a variable named spam has a type hint of int but is being assigned a float value. This could possibly cause a failure and should be investigated. Some error messages might be hard to understand at first reading. Mypy can report a large number of possible errors, too many to list here. The easiest way to find out what the error means is to search for it on the web. In this case, you might search for something like "Mypy incompatible types in assignment."

Running Mypy from the command line every time you change your code is rather inefficient. To make better use of a type checker, you'll need to configure your IDE or text editor to run it in the background. This way, the editor will constantly run Mypy as you type your code and then display any errors in the editor. Figure 11-1 shows the error from the previous example in the Sublime Text text editor.

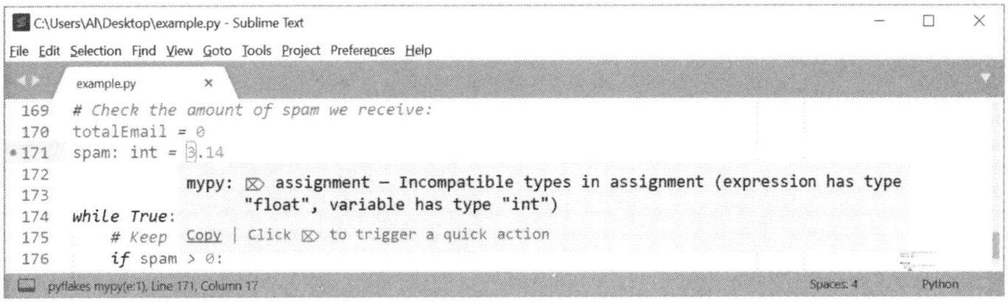

Figure 11-1: The Sublime Text text editor displaying errors from Mypy

The steps to configure your IDE or text editor to work with Mypy differ depending on which IDE or text editor you're using. You can find instructions online by searching for "*<your IDE>* Mypy configure," "*<your IDE>* type hints setup," or something similar. If all else fails, you can always run Mypy from the Command Prompt or Terminal window.

Telling Mypy to Ignore Code

You might write code that for whatever reason you don't want to receive type hint warnings about. To the static analysis tool, the line might appear to use the incorrect type, but it's actually fine when the program runs. You can suppress any type hint warnings by adding a # type: ignore comment to the end of the line. Here is an example:

```
def removeThreesAndFives(number: int) -> int:
    number = str(number)  # type: ignore
    number = number.replace('3', '').replace('5', '')  # type: ignore
    return int(number)
```

To remove all the 3 and 5 digits from the integer passed to removeThrees-AndFives(), we temporarily set the integer number variable to a string. This causes the type checker to warn us about the first two lines in the function, so we add the # type: ignore type hints to these lines to suppress the type checker's warnings.

Use # type: ignore sparingly. Ignoring warnings from the type checker provides an opening for bugs to sneak into your code. You can almost certainly rewrite your code so the warnings don't occur. For example, if we create a new variable with numberAsStr = str(number) or replace all three lines with a single return int(str(number.replace('3', '').replace('5', ''))) line of code, we can avoid reusing the number variable for multiple types. We wouldn't want to suppress the warning by changing the type hint for the parameter to Union[int, str], because the parameter is meant to allow integers only.

Setting Type Hints for Multiple Types

Python's variables, parameters, and return values can have multiple data types. To accommodate this, you can specify type hints with multiple types by importing Union from the built-in typing module. Specify a range of types inside square brackets following the Union class name:

```
from typing import Union
spam: Union[int, str, float] = 42
spam = 'hello'
spam = 3.14
```

In this example, the Union[int, str, float] type hint specifies that you can set spam to an integer, string, or floating-point number. Note that it's preferable to use the from typing import X form of the import statement rather than the import typing form and then consistently use the verbose typing.X for type hints throughout your program.

You might specify multiple data types in situations where a variable or return value could have the None value in addition to another type. To include NoneType, which is the type of the None value, in the type hint, place None inside the square brackets rather than NoneType. (Technically, NoneType isn't a built-in identifier the way int or str is.)

Better yet, instead of using, say, Union[str, None], you can import Optional from the typing module and use Optional[str]. This type hint means that the function or method could return None rather than a value of the expected type. Here's an example:

```
from typing import Optional
lastName: Optional[str] = None
lastName = 'Sweigart'
```

In this example, you could set the lastName variable to None or a str value. But it's best to make sparing use of Union and Optional. The fewer types your variables and functions allow, the simpler your code will be, and simple code is less bug prone than complicated code. Remember the Zen of Python maxim that simple is better than complex. For functions that return None to indicate an error, consider raising an exception instead. See "Raising Exceptions vs. Returning Error Codes" on page 178.

You can use the Any type hint (also from the typing module) to specify that a variable, parameter, or return value can be of any data type:

```
from typing import Any
import datetime
spam: Any = 42
spam = datetime.date.today()
spam = True
```

In this example, the Any type hint allows you to set the spam variable to a value of any data type, such as int, datetime.date, or bool. You can also use object as the type hint, because this is the base class for all data types in Python. But Any is a more readily understandable type hint than object.

As you should with Union and Optional, use Any sparingly. If you set all of your variables, parameters, and return values to the Any type hint, you'd lose the type-checking benefits of static typing. The difference between specifying the Any type hint and specifying no type hint is that Any explicitly states that the variable or function accepts values of any type, whereas an absent type hint indicates that the variable or function has yet to be type hinted.

Setting Type Hints for Lists, Dictionaries, and More

Lists, dictionaries, tuples, sets, and other container data types can hold other values. If you specify list as the type hint for a variable, that variable must contain a list, but the list could contain values of any type. The following code won't cause any complaints from a type checker:

```
spam: list = [42, 'hello', 3.14, True]
```

To specifically declare the data types of the values inside the list, you must use the typing module's List type hint. Note that List has a capital *L*, distinguishing it from the list data type:

```
from typing import List, Union
❶ catNames: List[str] = ['Zophie', 'Simon', 'Pooka', 'Theodore']
❷ numbers: List[Union[int, float]] = [42, 3.14, 99.9, 86]
```

In this example, the catNames variable contains a list of strings, so after importing List from the typing module, we set the type hint to List[str] ❶. The type checker catches any call to the append() or insert() method, or any other code that puts a nonstring value into the list. If the list should contain multiple types, we can set the type hint using Union. For example, the numbers list can contain integer and float values, so we set its type hint to List[Union[int, float]] ❷.

The typing module has a separate *type alias* for each container type. Here's a list of the type aliases for common container types in Python:

List is for the list data type.

Tuple is for the tuple data type.

Dict is for the dictionary (dict) data type.

Set is for the set data type.

FrozenSet is for the frozenset data type.

Sequence is for the list, tuple, and any other sequence data type.

Mapping is for the dictionary (dict), set, frozenset, and any other mapping data type.

ByteString is for the bytes, bytearray, and memoryview types.

You'll find the full list of these types online at *https://docs.python.org/3/library/typing.html#classes-functions-and-decorators*.

Backporting Type Hints with Comments

Backporting is the process of taking features from a new version of software and *porting* (that is, adapting and adding) them to an earlier version. Python's type hints feature is new to version 3.5. But in Python code that might be run by interpreter versions earlier than 3.5, you can still use type hints by putting the type information in comments. For variables, use an inline comment after the assignment statement. For functions and methods, write the type hint on the line following the def statement. Begin the comment with type:, followed by the data type. Here's an example of some code with type hints in the comments:

```
❶ from typing import List

❷ spam = 42  # type: int
  def sayHello():
    ❸ # type: () -> None
      """The docstring comes after the type hint comment."""
      print('Hello!')

  def addTwoNumbers(listOfNumbers, doubleTheSum):
    ❹ # type: (List[float], bool) -> float
      total = listOfNumbers[0] + listOfNumbers[1]
      if doubleTheSum:
          total *= 2
      return total
```

Note that even if you're using the comment type hint style, you still need to import the typing module ❶, as well as any type aliases that you use in the comments. Versions earlier than 3.5 won't have a typing module in their standard library, so you must install typing separately by running this command:

```
python -m pip install --user typing
```

Run python3 instead of python on macOS and Linux.

To set the spam variable to an integer, we add # type: int as the end-of-line comment ❷. For functions, the comment should include parentheses with a comma-separated list of type hints in the same order as the parameters. Functions with zero parameters would have an empty set of parentheses ❸. If there are multiple parameters, separate them inside the parentheses with commas ❹.

The comment type hint style is a bit less readable than the normal style, so use it only for code that might be run by versions of Python earlier than 3.5.

Summary

Programmers often forget about documenting their code. But by taking a little time to add comments, docstrings, and type hints to your code, you can avoid wasting time later. Well-documented code is also easier to maintain.

It's tempting to adopt the opinion that comments and documentation don't matter or are even a disadvantage when writing software. (Conveniently, this view allows programmers to avoid the work of writing documentation.) Don't be fooled; well-written documentation consistently saves you far more time and effort than it takes to write it. It's more common for programmers to stare at a screen of inscrutable, uncommented code than to have too much helpful information.

Good comments offer concise, useful, and accurate information to programmers who need to read the code at a later time and understand what it does. These comments should explain the original programmer's intent and summarize small sections of code rather than state the obvious about what a single line of code does. Comments sometimes offer detailed accounts of lessons the programmer learned while writing the code. This valuable information might spare future maintainers from having to relearn these lessons the hard way.

Docstrings, a Python-specific kind of comment, are multiline strings that appear immediately after the `class` or `def` statement, or at the top of the module. Documentation tools, such as Python's built-in `help()` function, can extract docstrings to provide specific information about the purpose of the class, function, or module.

Introduced in Python 3.5, type hints bring gradual typing to Python code. Gradual typing allows the programmer to apply the bug-detection benefits of static typing while maintaining the flexibility of dynamic typing. The Python interpreter ignores type hints, because Python doesn't have runtime type checking. Even so, static type-checking tools use type hints to analyze the source code when it isn't running. Type checkers, such as Mypy, can ensure that you aren't assigning invalid values to the variables you pass to functions. This saves you time and effort by preventing a broad category of bugs.

12

ORGANIZING YOUR CODE PROJECTS WITH GIT

Version control systems are tools that record all source code changes and make it easy to retrieve older versions of the code. Think of these tools as sophisticated *undo* features. For example, if you replace a function and then later decide you liked the old one better, you can restore your code to the original version. Or if you discover a new bug, you can go back to earlier versions to identify when it first appeared and which code change caused it.

A version control system manages files as you make changes to them. This is preferable to, say, making a copy of your *myProject* folder and naming it *myProject-copy*. If you keep making changes, you'll eventually have to make another copy named *myProject-copy2*, then *myProject-copy3*, *myProject-copy3b*, *myProject-copyAsOfWednesday*, and so on. Copying folders might be simple, but the approach doesn't scale. Learning to use a version control system saves you time and headaches in the long run.

Git, Mercurial, and Subversion are popular version control applications, although Git is by far the most popular. In this chapter, you'll learn how to set up files for code projects and use Git to track their changes.

Git Commits and Repos

Git allows you to save the state of your project files, called *snapshots* or *commits*, as you make changes to them. That way, you can roll back to any previous snapshot if you ever need to. Commit is a noun and a verb; programmers commit (or save) their commits (or snapshots). *Check-in* is also a less popular term for commits.

Version control systems also make it easy for a software developer team to remain in sync with each other while they make changes to a project's source code. As each programmer commits their changes, other programmers can pull these updates onto their computers. The version control system tracks what commits were made, who made them, and when they made them, along with the developers' comments describing the changes.

Version control manages a project's source code in a folder called a repository, or *repo*. In general, you should keep a separate Git repo for each project you're working on. This chapter assumes you're mostly working on your own and don't need the advanced Git features, such as branching and merging, that help programmers collaborate. But even if you're working alone, there is no programming project too small to benefit from version control.

Using Cookiecutter to Create New Python Projects

We call the folder that contains all the source code, documentation, tests, and other files related to a project the *working directory* or *working tree* in Git parlance, and *project folder* more generally. The files in the working directory are collectively called the *working copy*. Before we create our Git repo, let's create the files for a Python project.

Every programmer has a preferred method for doing so. Even so, Python projects follow conventions for folder names and hierarchies. Your simpler programs might consist of a single *.py* file. But as you tackle more sophisticated projects, you'll start to include additional *.py* files, data files, documentation, unit tests, and more. Typically, the root of the project folder contains a *src* folder for the *.py* source code files, a *tests* folder for unit tests, and a *docs* folder for any documentation (such as those generated by the Sphinx documentation tool). Other files contain project information and tool configuration: *README.md* for general information, *.coveragerc* for code coverage configuration, *LICENSE.txt* for the project's software license, and so on. These tools and files are beyond the scope of this book, but they're worth investigating. As you gain more coding experience, re-creating the same basic files for new programming projects becomes tedious. To speed up your coding tasks, you can use the cookiecutter Python module to create these files and folders automatically. You'll find the full documentation for both the module and the Cookiecutter command line program at *https:// cookiecutter.readthedocs.io/.*

To install Cookiecutter, run `pip install --user cookiecutter` (on Windows) or `pip3 install --user cookiecutter` (on macOS and Linux). This installation includes the Cookiecutter command line program and the cookiecutter Python module. The output might warn you that the command line program is installed to a folder not listed in the `PATH` environment variable:

```
Installing collected packages: cookiecutter
  WARNING: The script cookiecutter.exe is installed in 'C:\Users\Al\AppData\
Roaming\Python\Python38\Scripts' which is not on PATH.
  Consider adding this directory to PATH or, if you prefer to suppress this
warning, use --no-warn-script-location.
```

Consider adding the folder (*C:\Users\Al\AppData\Roaming\Python\ Python38\Scripts* in this case) to the `PATH` environment variable by following the instructions in "Environment Variables and PATH" on page 35. Otherwise, you'll have to run Cookiecutter as a Python module by entering `python -m cookiecutter` (on Windows) or `python3 -m cookiecutter` (on macOS and Linux) instead of simply `cookiecutter`.

In this chapter, we'll create a repo for a module named `wizcoin`, which handles the galleon, sickle, and knut coins of a fictional wizarding currency. The cookiecutter module uses templates to create the starting files for several different kinds of projects. Often, the template is simply a *GitHub.com* link. For example, from a *C:\Users\Al* folder, you could enter the following in a Terminal window to create a *C:\Users\Al\wizcoin* folder with the boilerplate files for a basic Python project. The cookiecutter module downloads the template from GitHub and asks you a series of questions about the project you want to create:

```
C:\Users\Al>cookiecutter gh:asweigart/cookiecutter-basicpythonproject
project_name [Basic Python Project]: WizCoin
module_name [basicpythonproject]: wizcoin
author_name [Susie Softwaredeveloper]: Al Sweigart
author_email [susie@example.com]: al@inventwithpython.com
github_username [susieexample]: asweigart
project_version [0.1.0]:
project_short_description [A basic Python project.]: A Python module to
represent the galleon, sickle, and knut coins of wizard currency.
```

If you get an error, you can also run `python -m cookiecutter` instead of cookiecutter. This command downloads a template I've created from *https:// github.com/asweigart/cookiecutter-basicpythonproject.* You'll find templates for many programming languages at *https://github.com/cookiecutter/cookiecutter.* Because Cookiecutter templates are often hosted on GitHub, you could also enter gh: as a shortcut for `https://github.com/` in the command line argument.

As Cookiecutter asks you questions, you can either enter a response or simply press ENTER to use the default response shown in between square brackets. For example, `project_name [Basic Python Project]:` asks you to name your project. If you enter nothing, Cookiecutter will use "Basic Python Project" as the project name. These defaults also hint at what sort of response is expected. The `project_name [Basic Python Project]:` prompt shows you a capitalized project name that includes spaces, whereas the `module_name [basicpythonproject]:` prompt shows you that the module name is lowercase and has no spaces. We didn't enter a response for the `project_version [0.1.0]:` prompt, so the response defaults to "0.1.0."

After answering the questions, Cookiecutter creates a *wizcoin* folder in the current working directory with the basic files you'll need for a Python project, as shown in Figure 12-1.

Name	Date modified	Type	Size
docs	8/31/2021 12:37 PM	File folder	
src	8/31/2021 12:37 PM	File folder	
tests	8/31/2021 12:37 PM	File folder	
.coveragerc	8/31/2021 12:37 PM	COVERAGERC File	1 KB
.gitignore	8/31/2021 12:37 PM	Text Document	2 KB
code_of_conduct....	8/31/2021 12:37 PM	MD File	4 KB
LICENSE.txt	8/31/2021 12:37 PM	TXT File	35 KB
pyproject.toml	8/31/2021 12:37 PM	TOML File	0 KB
README.md	8/31/2021 12:37 PM	MD File	1 KB
setup.py	8/31/2021 12:37 PM	PY File	2 KB
tox.ini	8/31/2021 12:37 PM	INI File	1 KB

Figure 12-1: The files in the wizcoin *folder created by Cookiecutter*

It's okay if you don't understand the purpose of these files. A full explanation of each is beyond the scope of this book, but *https://github.com/asweigart/cookiecutter-basicpythonproject* has links and descriptions for further reading. Now that we have our starting files, let's keep track of them using Git.

Installing Git

Git might already be installed on your computer. To find out, run `git --version` from the command line. If you see a message like `git version 2.29.0.windows.1`, you already have Git. If you see a "command not found" error message, you must install Git. On Windows, go to *https://git-scm .com/download*, and then download and run the Git installer. On macOS Mavericks (10.9) and later, simply run `git --version` from the terminal and you'll be prompted to install Git, as shown in Figure 12-2.

On Ubuntu or Debian Linux, run `sudo apt install git-all` from the terminal. On Red Hat Linux, run `sudo dnf install git-all` from the terminal. Find instructions for other Linux distributions at *https://git-scm.com/download/ linux*. Confirm that the install worked by running `git --version`.

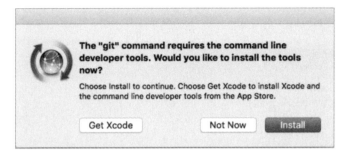

Figure 12-2: The first time you run git --version on macOS 10.9 or later, you'll be prompted to install Git.

Configuring Your Git Username and Email

After installing Git, you need to configure your name and email so your commits include your author information. From a terminal, run the following git config commands using your name and email information:

```
C:\Users\Al>git config --global user.name "Al Sweigart"
C:\Users\Al>git config --global user.email al@inventwithpython.com
```

This configuration information is stored in a *.gitconfig* file in your home folder (such as *C:\Users\Al* on my Windows laptop). You'll never need to edit this text file directly. Instead, you can change it by running the git config command. You can list the current Git configuration settings using the git config --list command.

Installing GUI Git Tools

This chapter focuses on the Git command line tool, but installing software that adds a GUI for Git can help you with day-to-day tasks. Even professional software developers who know the CLI Git commands often use GUI Git tools. The web page at *https://git-scm.com/downloads/guis* suggests several of these tools, such as TortoiseGit for Windows, GitHub Desktop for macOS, and GitExtensions for Linux.

For example, Figure 12-3 shows how TortoiseGit on Windows adds overlays to File Explorer's icons based on their status: green for unmodified repo files, red for modified repo files (or folders containing modified files), and no icon for untracked files. Checking these overlays is certainly more convenient than constantly entering commands into a terminal for this information. TortoiseGit also adds a context menu for running Git commands, as shown in Figure 12-3.

Using GUI Git tools is convenient, but it's not a substitute for learning the command line commands featured in this chapter. Keep in mind that you might need to one day use Git on a computer that doesn't have these GUI tools installed.

Figure 12-3: TortoiseGit for Windows adds a GUI to run Git commands from File Explorer.

The Git Workflow

Using a Git repo involves the following steps. First, you create the Git repo by running the git init or git clone command. Second, you add files with the git add <filename> command for the repo to track. Third, once you've added files, you can commit them with the git commit -am "<descriptive commit message>" command. At this point, you're ready to make more changes to your code.

You can view the help file for each of these commands by running git help <command>, such as git help init or git help add. These help pages are handy for reference, although they're too dry and technical to use as tutorials. You'll learn more details about each of these commands later, but first, you need to understand a few Git concepts to make the rest of this chapter easier to digest.

How Git Keeps Track of File Status

All files in a working directory are either tracked or untracked by Git. *Tracked files* are the files that have been added and committed to the repo, whereas every other file is *untracked*. To the Git repo, untracked files in the working copy might as well not exist. On the other hand, the tracked files exist in one of three other states:

- The *committed state* is when a file in the working copy is identical to the repo's most recent commit. (This is also sometimes called the *unmodified state* or *clean state*.)
- The *modified state* is when a file in the working copy is different than the repo's most recent commit.

- The *staged state* is when a file has been modified and marked to be included in the next commit. We say that the file is *staged* or *in the staging area*. (The staging area is also known as the *index* or *cache*.)

Figure 12-4 contains a diagram of how a file moves between these four states. You can add an untracked file to the Git repo, in which case it becomes tracked and staged. You can then commit staged files to put them into the committed state. You don't need any Git command to put a file into the modified state; once you make changes to a committed file, it's automatically labeled as modified.

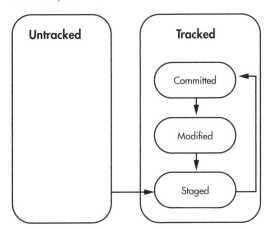

Figure 12-4: The possible states of a file in a Git repo and the transitions between them

At any step after you've created the repo, run git status to view the current status of the repo and its files' states. You'll frequently run this command as you work in Git. In the following example, I've set up files in different states. Notice how these four files appear in the output of git status:

```
C:\Users\Al\ExampleRepo>git status
On branch master
Changes to be committed:
  (use "git restore --staged <file>..." to unstage)
    ❶ new file:   new_file.py
    ❷ modified:   staged_file.py

Changes not staged for commit:
  (use "git add <file>..." to update what will be committed)
  (use "git restore <file>..." to discard changes in working directory)
    ❸ modified:   modified_file.py

Untracked files:
  (use "git add <file>..." to include in what will be committed)
    ❹ untracked_file.py
```

In this working copy, there's a *new_file.py* ❶, which has recently been added to the repo and is therefore in the staged state. There are also two tracked files, *staged_file.py* ❷ and *modified_file.py* ❸, which are in the staged and modified states, respectively. Then there's an untracked file named *untracked_file.py* ❹. The output of git status also has reminders for the Git commands that move the files to other states.

Why Stage Files?

You might wonder what the point of the staged state is. Why not just go between *modified* and *committed* without staging files? Dealing with the staging area is full of thorny special cases and a large source of confusion for Git beginners. For instance, a file can be modified after it has been staged, leading to files existing in both the modified and staged states, as described in the previous section. Technically, the staging area doesn't contain files so much as changes, because parts of a single modified file can be staged and other parts unstaged. Cases like these are why Git has a reputation for being complex, and many sources of information on how Git works are often imprecise at best and misleading at worst.

But we can avoid most of this complexity. In this chapter, I recommend avoiding it by using the git commit -am command to stage and commit modified files in a single step. This way they'll move directly from the modified state to the clean state. Also, I recommend always immediately committing files after adding, renaming, or removing them in your repo. Additionally, using GUI Git tools (explained later) rather than the command line can help you avoid these tricky cases.

Creating a Git Repo on Your Computer

Git is a *distributed version control system*, which means it stores all of its snapshots and repo metadata locally on your computer in a folder named *.git*. Unlike a centralized version control system, Git doesn't need to connect to a server over the internet to make commits. This makes Git fast and available to work with when you're offline.

From a terminal, run the following commands to create the *.git* folder. (On macOS and Linux, you'll need to run mkdir instead of md.)

```
C:\Users\Al>md wizcoin
C:\Users\Al>cd wizcoin
C:\Users\Al\wizcoin>git init
Initialized empty Git repository in C:/Users/Al/wizcoin/.git/
```

When you convert a folder into a Git repo by running git init, all the files in it start as untracked. For our *wizcoin* folder, the git init command creates the *wizcoin/.git* folder, which contains the Git repo metadata. The presence of this *.git* folder makes a folder a Git repository; without it, you simply have a collection of source code files in an ordinary folder. You'll never have to directly modify the files in *.git*, so just ignore this folder. In fact, it's named *.git* because most operating systems automatically hide folders and files whose names begin with a period.

Now you have a repo in your *C:\Users\Al\wizcoin* working directory. A repo on your computer is known as a *local repo*; a repo located on someone else's computer is known as a *remote repo*. This distinction is important, because you'll often have to share commits between local and remote repos so you can work with other developers on the same project.

You can now use the git command to add files and track changes within the working directory. If you run git status in your newly created repo, you'll see the following:

```
C:\Users\Al\wizcoin>git status
On branch master

No commits yet

nothing to commit (create/copy files and use "git add" to track)
```

The output from this command informs you that you have no commits yet in this repo.

RUNNING GIT STATUS WITH THE WATCH COMMAND

While using the Git command line tool, you'll often run git status to see your repo's status. Instead of entering this command manually, you can use the watch command to run it for you. The watch command runs a given command repeatedly every two seconds, refreshing the screen with its latest output.

On Windows, you can obtain the watch command by downloading *https://inventwithpython.com/watch.exe* and placing this file into a PATH folder, such as *C:\Windows*. On macOS, you can go to *https://www.macports.org/* to download and install MacPorts, and then run sudo ports install watch. On Linux, the watch command is already installed. Once it's installed, open a new Command Prompt or Terminal window, run **cd** to change directory to your Git repo's project folder, and run **watch "git status"**. The watch command will run git status every two seconds, displaying the latest results onscreen. You can leave this window open while you use the Git command line tool in a different Terminal window to see how your repo's status changes in real time. You can open another Terminal window and run watch "git log –online" to view a summary of the commits you make, also updated in real time. This information helps remove the guesswork as to what the Git commands you enter are doing to your repo.

Adding Files for Git to Track

Only tracked files can be committed, rolled back, or otherwise interacted with through the git command. Run **git status** to see the status of the files in the project folder:

```
C:\Users\Al\wizcoin>git status
On branch master

No commits yet
```
❶ ```
Untracked files:
  (use "git add <file>..." to include in what will be committed)

        .coveragerc
        .gitignore
        LICENSE.txt
        README.md
--snip--
        tox.ini

nothing added to commit but untracked files present (use "git add" to track)
```

All the files in the *wizcoin* folder are currently untracked ❶. We can track them by doing an initial commit of these files, which takes two steps: running git add for each file to be committed, and then running git commit to create a commit of all these files. Once you've committed a file, Git tracks it.

The git add command moves files from the untracked state or modified state to the staged state. We could run git add for every file we plan to stage (for example, git add .coveragerc, git add .gitignore, git add LICENSE.txt, and so on), but that's tedious. Instead, let's use the * wildcard to add several files at once. For example, git add *.py adds all *.py* files in the current working directory and its subdirectories. To add every untracked file, use a single period (.) to tell Git to match all files:

```
C:\Users\Al\wizcoin>git add .
```

Run git status to see the files you've staged:

```
C:\Users\Al\wizcoin>git status
On branch master

No commits yet
```
❶ ```
Changes to be committed:
  (use "git rm --cached <file>..." to unstage)

    ❷ new file: .coveragerc
        new file: .gitignore
--snip--
        new file: tox.ini
```

The output of git status tells you which files are staged to be commit-ted the next time you run git commit ❶. It also tells you that these are new files added to the repo ❷ rather than existing files in the repo that have been modified.

After running git add to select the files to add to the repo, run git commit -m "Adding new files to the repo." (or a similar commit message) and git status again to view the repo status:

```
C:\Users\Al\wizcoin>git commit -m "Adding new files to the repo."
[master (root-commit) 65f3b4d] Adding new files to the repo.
 15 files changed, 597 insertions(+)
 create mode 100644 .coveragerc
 create mode 100644 .gitignore
--snip--
 create mode 100644 tox.ini

C:\Users\Al\wizcoin>git status
On branch master
nothing to commit, working tree clean
```

Note that any files listed in the *.gitignore* file won't be added to staging, as I explain in the next section.

Ignoring Files in the Repo

Files not tracked by Git appear as untracked when you run git status. But in the course of writing your code, you might want to exclude certain files from version control completely so you don't accidentally track them. These include:

- Temporary files in the project folder
- The *.pyc*, *.pyo*, and *.pyd* files that the Python interpreter generates when it runs *.py* programs
- The *.tox*, *htmlcov*, and other folders that various software development tools generate *docs/_build*
- Any other compiled or generated files that could be regenerated (because the repo is for source files, not the products created from source files)
- Source code files that contain database passwords, authentication tokens, credit card numbers, or other sensitive information

To avoid including these files, create a text file named *.gitignore* that lists the folders and files that Git should never track. Git will automatically exclude these from git add or git commit commands, and they won't appear when you run git status.

The *.gitignore* file that the cookiecutter-basicpythonproject template creates looks like this:

```
# Byte-compiled / optimized / DLL files
__pycache__/
*.py[cod]
*$py.class
--snip--
```

The *.gitignore* file uses * for wildcards and # for comments. You can read more about it in the online documentation at *https://git-scm.com/docs/gitignore*.

You should add the actual *.gitignore* file to the Git repo so other programmers have it if they clone your repo. If you want to see which files in your working directory are being ignored based on the settings in *.gitignore*, run the git ls-files --other --ignored --exclude-standard command.

Committing Changes

After adding new files to the repo, you can continue writing code for your project. When you want to create another snapshot, you can run git add . to stage all modified files and git commit -m *<commit message>* to commit all staged files. But doing so is easier with the single git commit -am *<commit message>* command:

```
C:\Users\Al\wizcoin>git commit -am "Fixed the currency conversion bug."
[master (root-commit) e1ae3a3] Fixed the currency conversion bug.
 1 file changed, 12 insertions(+)
```

If you want to commit only certain modified files instead of every modified file, you can omit the –a option from –am and specify the files after the commit message, such as git commit -m *<commit message>* file1.py file2.py.

The commit message provides a hint for future use: it's a reminder about what changes we made in this commit. It might be tempting to write a short, generic message, such as "Updated code," or "Fixed a few bugs," or even just "x" (because blank commit messages aren't allowed). But three weeks from now, when you need to roll back to an earlier version of your code, detailed commit messages will save you a lot of grief in determining exactly how far back you need to go.

If you forget to add the -m "*<message>*" command line argument, Git will open the Vim text editor in the Terminal window. Vim is beyond the scope of this book, so press the ESC key and enter qa! to safely exit Vim and cancel the commit. Then enter the git commit command again, this time with the -m "*<message>*" command line argument.

For examples of what professional commit messages look like, check out the commit history for the Django web framework at *https://github.com/django/django/commits/master*. Because Django is a large, open source project, the commits occur frequently and are formal commit messages. Infrequent commits with vague commit messages might work well enough for your small, personal programming projects, but Django has more than 1,000 contributors. Poor commit messages from any of them becomes a problem for all of them.

The files are now safely committed to the Git repo. Run git status one more time to view their status:

```
C:\Users\Al\wizcoin>git status
On branch master
nothing to commit, working tree clean
```

By committing the staged files, you've moved them back to the committed state, and Git tells us that the working tree is clean; in other words, there are no modified or staged files. To recap, when we added files to the Git repo, the files went from untracked to staged and then to committed. The files are now ready for future modifications.

Note that you can't commit folders to a Git repo. Git automatically includes folders in the repo when a file in them is committed, but you can't commit an empty folder.

If you made a typo in the most recent commit message, you can rewrite it using the git commit --amend -m "<new commit message>" command.

Using git diff to View Changes Before Committing

Before you commit code, you should quickly review the changes you'll commit when you run git commit. You can view the differences between the code currently in your working copy and the code in the latest commit using the git diff command.

Let's walk through an example of using git diff. Open *README.md* in a text editor or IDE. (You should have created this file when you ran Cookiecutter. If it doesn't exist, create a blank text file and save it as *README.md*.) This is a Markdown-formatted file, but like Python scripts, it's written in plaintext. Change the TODO - fill this in later text in the Quickstart Guide section to the following (keep the xample typo in it for now; we'll fix it later):

```
Quickstart Guide
----------------

Here's some xample code demonstrating how this module is used:

    >>> import wizcoin
    >>> coin = wizcoin.WizCoin(2, 5, 10)
    >>> str(coin)
    '2g, 5s, 10k'
    >>> coin.value()
    1141
```

Before we add and commit *README.md*, run the git diff command to see the changes we've made:

```
C:\Users\Al\wizcoin>git diff
diff --git a/README.md b/README.md
```

```
index 76b5814..3be49c3 100644
--- a/README.md
+++ b/README.md
@@ -13,7 +13,14 @@ To install with pip, run:
 Quickstart Guide
 ----------------

-TODO - fill this in later
+Here's some xample code demonstrating how this module is used:
+
+    >>> import wizcoin
+    >>> coin = wizcoin.WizCoin(2, 5, 10)
+    >>> str(coin)
+    '2g, 5s, 10k'
+    >>> coin.value()
+    1141

 Contribute
 ----------
```

The output shows that *README.md* in your working copy has changed from the *README.md* as it exists in the latest commit of the repo. The lines that begin with a minus sign - have been removed; the lines that begin with a plus sign + have been added.

While reviewing the changes, you'll also notice that we made a typo by writing xample instead of example. We shouldn't check in this typo. Let's correct it. Then run git diff again to inspect the change and add and commit it to the repo:

```
C:\Users\Al\wizcoin>git diff
diff --git a/README.md b/README.md
index 76b5814..3be49c3 100644
--- a/README.md
+++ b/README.md
@@ -13,7 +13,14 @@ To install with pip, run:
 Quickstart Guide
 ----------------

-TODO - fill this in later
+Here's some example code demonstrating how this module is used:
--snip--
C:\Users\Al\wizcoin>git add README.md

C:\Users\Al\wizcoin>git commit -m "Added example code to README.md"
[master 2a4c5b8] Added example code to README.md
 1 file changed, 8 insertions(+), 1 deletion(-)
```

The correction is now safely committed to the repo.

Using git difftool to View Changes with a GUI Application

It's easier to see changes with a diff program that uses a GUI. On Windows, you can download WinMerge (*https://winmerge.org/*), a free, open source diff program, and then install it. On Linux, you can install either Meld by using the `sudo apt-get install meld` command or Kompare by using the `sudo apt-get install kompare` command. On macOS, you can install tkdiff by using commands that first install and configure Homebrew (a package manager that installs software) and then using Homebrew to install tkdiff:

```
/bin/bash -c "$(curl -fsSL https://raw.githubusercontent.com/Homebrew/install/
master/install.sh)"
brew install tkdiff
```

You can configure Git to use these tools by running `git config diff.tool` *<tool_name>*, where *<tool_name>* is winmerge, tkdiff, meld, or kompare. Then run `git difftool` *<filename>* to view the changes made to a file in the tool's GUI, as shown in Figure 12-5.

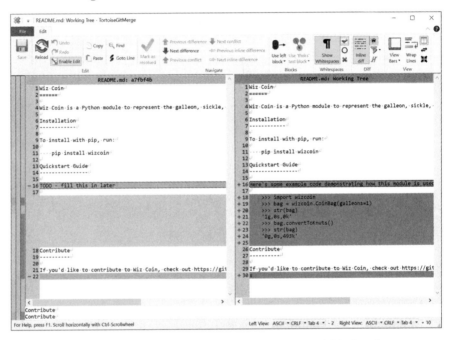

Figure 12-5: A GUI diff tool, in this case WinMerge, is more readable than the text output of git diff.

Additionally, run `git config --global difftool.prompt false` so Git doesn't ask for confirmation each time you want to open the diff tool. If you installed a GUI Git client, you can also configure it to use these tools (or it might come with a visual diff tool of its own).

How Often Should I Commit Changes?

Even though version control allows you to roll back your files to an earlier commit, you might wonder how often you should make commits. If you commit too frequently, you'll have trouble sorting through a large number of insignificant commits to find the version of the code you're looking for.

If you commit too infrequently, each commit will contain a large number of changes, and reverting to a particular commit will undo more changes than you want to. In general, programmers tend to commit less frequently than they should.

You should commit code when you've completed an entire piece of functionality, such as a feature, class, or bug fix. Don't commit any code that contains syntax errors or is obviously broken. Commits can consist of a few lines of changed code or several hundred, but either way, you should be able to jump back to any earlier commit and still have a working program. You should always run any unit tests before committing. Ideally, all your tests should pass (and if they don't pass, mention this in the commit message).

Deleting Files from the Repo

If you no longer need Git to track a file, you can't simply delete the file from the filesystem. You must delete it through Git using the git rm command, which also tells Git to untrack the file. To practice doing so, run the echo "Test file" > deleteme.txt command to create a small file named *deleteme. txt* with the contents "Test file". Then commit it to the repo by running the following commands:

```
C:\Users\Al\wizcoin>echo "Test file" > deleteme.txt
C:\Users\Al\wizcoin>git add deleteme.txt
C:\Users\Al\wizcoin>git commit -m "Adding a file to test Git deletion."
[master 441556a] Adding a file to test Git deletion.
 1 file changed, 1 insertion(+)
 create mode 100644 deleteme.txt
C:\Users\Al\wizcoin>git status
On branch master
nothing to commit, working tree clean
```

Don't delete the file using the del command on Windows or rm command on macOS and Linux. (If you do, you can run git restore *<filename>* to recover it or simply continue with the git rm command to remove it from the repo.) Instead, use the git rm command to delete and stage the *deleteme.txt* file such as in this example:

```
C:\Users\Al\wizcoin>git rm deleteme.txt
rm deleteme.txt'
```

The git rm command deletes the file from your working copy, but you're not done yet. Like git add, the git rm command stages the file. You need to commit file deletion just like any other change:

```
C:\Users\Al\wizcoin>git status
On branch master
Changes to be committed:
❶ (use "git reset HEAD <file>..." to unstage)

        deleted:    deleteme.txt
```

```
C:\Users\Al\wizcoin>git commit -m "Deleting deleteme.txt from the repo to
finish the deletion test."
[master 369de78] Deleting deleteme.txt from the repo to finish the deletion
test.
 1 file changed, 1 deletion(-)
 delete mode 100644 deleteme.txt
C:\Users\Al\Desktop\wizcoin>git status
On branch master
nothing to commit, working tree clean
```

Even though you've deleted *deleteme.txt* from your working copy, it still exists in the repo's history. The "Recovering Old Changes" section later in this chapter describes how to recover a deleted file or undo a change.

The git rm command only works on files that are in the clean, committed state, without any modifications. Otherwise, Git asks you to commit the changes or revert them with the git reset HEAD *<filename>* command. (The output of git status reminds you of this command ❶.) This procedure prevents you from accidentally deleting uncommitted changes.

Renaming and Moving Files in the Repo

Similar to deleting a file, you shouldn't rename or move a file in a repo unless you use Git. If you try to do so without using Git, it will think you deleted a file and then created a new file that just happens to have the same content. Instead, use the git mv command, followed by git commit. Let's rename the *README.md* file to *README.txt* by running the following commands:

```
C:\Users\Al\wizcoin>git mv README.md README.txt
C:\Users\Al\wizcoin>git status
On branch master
Changes to be committed:
  (use "git reset HEAD <file>..." to unstage)

        renamed:    README.md -> README.txt


C:\Users\Al\wizcoin>git commit -m "Testing the renaming of files in Git."
[master 3fee6a6] Testing the renaming of files in Git.
 1 file changed, 0 insertions(+), 0 deletions(-)
 rename README.md => README.txt (100%)
```

This way, the history of changes to *README.txt* also includes the history of *README.md*.

We can also use the git mv command to move a file to a new folder. Enter the following commands to create a new folder called *movetest* and move the *README.txt* into it:

```
C:\Users\Al\wizcoin>mkdir movetest
C:\Users\Al\wizcoin>git mv README.txt movetest/README.txt
C:\Users\Al\wizcoin>git status
On branch master
```

```
Changes to be committed:
  (use "git reset HEAD <file>..." to unstage)

        renamed:    README.txt -> movetest/README.txt

C:\Users\Al\wizcoin>git commit -m "Testing the moving of files in Git."
[master 3ed22ed] Testing the moving of files in Git.
 1 file changed, 0 insertions(+), 0 deletions(-)
 rename README.txt => movetest/README.txt (100%)
```

You can also rename and move a file at the same time by passing git mv
a new name and location. Let's move the *README.txt* back to its original
place at the root of the working directory and give it its original name:

```
C:\Users\Al\wizcoin>git mv movetest/README.txt README.md
C:\Users\Al\wizcoin>git status
On branch master
Changes to be committed:
  (use "git reset HEAD <file>..." to unstage)

        renamed:    movetest/README.txt -> README.md

C:\Users\Al\wizcoin>git commit -m "Moving the README file back to its original
place and name."
[master 962a8ba] Moving the README file back to its original place and name.
 1 file changed, 0 insertions(+), 0 deletions(-)
 rename movetest/README.txt => README.md (100%)
```

Note that even though the *README.md* file is back in its original folder
and has its original name, the Git repo remembers the moves and name
changes. You can see this history using the git log command, described in
the next section.

Viewing the Commit Log

The git log command outputs a list of all commits:

```
C:\Users\Al\wizcoin>git log
commit 962a8baa29e452c74d40075d92b00897b02668fb (HEAD -> master)
Author: Al Sweigart <al@inventwithpython.com>
Date:   Wed Sep 1 10:38:23 2021 -0700

    Moving the README file back to its original place and name.

commit 3ed22ed7ae26220bbd4c4f6bc52f4700dbb7c1f1
Author: Al Sweigart <al@inventwithpython.com>
Date:   Wed Sep 1 10:36:29 2021 -0700

    Testing the moving of files in Git.

--snip--
```

This command can display a large amount of text. If the log won't fit in your Terminal window, it'll let you scroll up or down using the up and down arrow keys. To quit, press the **q** key.

If you want to set your files to a commit that's earlier than the latest one, you need to first find the *commit hash*, a 40-character string of hexadecimal digits (composed of numbers and the letters A to F), which works as a unique identifier for a commit. For example, the full hash for the most recent commit in our repo is 962a8baa29e452c74d40075d92b00897b02668fb. But it's common to use only the first seven digits: 962a8ba.

Over time, the log can get very lengthy. The --oneline option trims the output to abbreviated commit hashes and the first line of each commit message. Enter git log --oneline into the command line:

```
C:\Users\Al\wizcoin>git log --oneline
962a8ba (HEAD -> master) Moving the README file back to its original place and
name.
3ed22ed Testing the moving of files in Git.
15734e5 Deleting deleteme.txt from the repo to finish the deletion test.
441556a Adding a file to test Git deletion.
2a4c5b8 Added example code to README.md
e1ae3a3 An initial add of the project files.
```

If this log is still too long, you can use -n to limit the output to the most recent commits. Try entering git log --oneline -n 3 to view only the last three commits:

```
C:\Users\Al\wizcoin>git log --oneline -n 3
962a8ba (HEAD -> master) Moving the README file back to its original place and
name.
3ed22ed Testing the moving of files in Git.
15734e5 Deleting deleteme.txt from the repo to finish the deletion test.
```

To display the contents of a file as it was at a particular commit, you can run the git show *<hash>*:*<filename>* command. But GUI Git tools will provide a more convenient interface for examining the repo log than the command line Git tool provides.

Recovering Old Changes

Let's say you want to work with an earlier version of your source code because you've introduced a bug, or perhaps you accidentally deleted a file. A version control system lets you undo, or *roll back*, your working copy to the content of an earlier commit. The exact command you'll use depends on the state of the files in the working copy.

Keep in mind that version control systems only add information. Even when you delete a file from a repo, Git will remember it so you can restore it later. Rolling back a change actually adds a new change that sets a file's content to its state in a previous commit. You'll find detailed information on various kinds of rollbacks at *https://github.blog/2015-06-08-how-to-undo-almost-anything-with-git/*.

Undoing Uncommitted Local Changes

If you've made uncommitted changes to a file but want to revert it to the version in the latest commit, you can run git restore *<filename>*. In the following example, we modify the *README.md* file but don't yet stage or commit it:

```
C:\Users\Al\wizcoin>git status
On branch master

Changes not staged for commit:
  (use "git add <file>..." to update what will be committed)
  (use "git restore <file>..." to discard changes in working directory)
        modified:   README.md

no changes added to commit (use "git add" and/or "git commit -a")
C:\Users\Al\wizcoin>git restore README.md
C:\Users\Al\wizcoin>git status
On branch master
Your branch is up to date with 'origin/master'.

nothing to commit, working tree clean
```

After you've run the git restore README.md command, the content of *README.md* reverts to that of the last commit. This is effectively an *undo* for the changes you've made to the file (but haven't yet staged or committed). But be careful: you can't undo this "undo" to get those changes back.

You can also run git checkout . to revert all changes you've made to every file in your working copy.

Unstaging a Staged File

If you've staged a modified file by running the git add command on it but now want to remove it from staging so it won't be included in the next commit, run git restore --staged *<filename>* to unstage it:

```
C:\Users\Al>git restore --staged README.md
Unstaged changes after reset:
M       spam.txt
```

README.md remains modified as it was before git add staged the file, but the file is no longer in the staged state.

Rolling Back the Most Recent Commits

Suppose you've made several unhelpful commits and you want to start over from a previous commit. To undo a specific number of the most recent commits, say, three, use the git revert -n HEAD~3..HEAD command. You can replace the 3 with any number of commits. For example, let's say you tracked the changes to a mystery novel you were writing and have the following Git log of all your commits and commit messages.

```
C:\Users\Al\novel>git log --oneline
de24642 (HEAD -> master) Changed the setting to outer space.
2be4163 Added a whacky sidekick.
97c655e Renamed the detective to 'Snuggles'.
8aa5222 Added an exciting plot twist.
2590860 Finished chapter 1.
2dece36 Started my novel.
```

Later you decide you want to start over again from the exciting plot twist at hash 8aa5222. This means you should undo the changes from the last three commits: de24642, 2be4163, and 97c655e. Run git revert -n HEAD~3..HEAD to undo these changes, and then run git add . and git commit -m "<*commit message*>" to commit this content, just as you would with any other change:

```
C:\Users\Al\novel>git revert -n HEAD~3..HEAD

C:\Users\Al\novel>git add .

C:\Users\Al\novel>git commit -m "Starting over from the plot twist."
[master faec20e] Starting over from the plot twist.
 1 file changed, 34 deletions(-)

C:\Users\Al\novel>git log --oneline
faec20e (HEAD -> master) Starting over from the plot twist.
de24642 Changed the setting to outer space.
2be4163 Added a whacky sidekick.
97c655e Renamed the detective to 'Snuggles'.
8aa5222 Added an exciting plot twist.
2590860 Finished chapter 1.
2dece36 Started my novel.
```

Git repos generally only add information, so undoing these commits still leaves them in the commit history. If you ever want to undo this "undo," you can roll it back using git revert again.

Rolling Back to a Specific Commit for a Single File

Because commits capture the state of the entire repo instead of individual files, you'll need a different command if you want to roll back changes for a single file. For example, let's say I had a Git repo for a small software project. I've created an *eggs.py* file and added functions spam() and bacon(), and then renamed bacon() to cheese(). The log for this repo would look something like this:

```
C:\Users\Al\myproject>git log --oneline
895d220 (HEAD -> master) Adding email support to cheese().
df617da Renaming bacon() to cheese().
ef1e4bb Refactoring bacon().
ac27c9e Adding bacon() function.
009b7c0 Adding better documentation to spam().
0657588 Creating spam() function.
d811971 Initial add.
```

But I've decided I want to revert the file back to before I added bacon() without changing any other files in the repo. I can use the git show *<hash>*: *<filename>* command to display this file as it was after a specific commit. The command would look something like this:

```
C:\Users\Al\myproject>git show 009b7c0:eggs.py
<contents of eggs.py as it was at the 009b7c0 commit>
```

Using the git checkout *<hash>* -- *<filename>*, I could set the contents of *eggs.py* to this version and commit the changed file as normal. The git checkout command only changes the working copy. You'll still need to stage and commit these changes like any other change:

```
C:\Users\Al\myproject>git checkout 009b7c0 -- eggs.py

C:\Users\Al\myproject>git add eggs.py

C:\Users\Al\myproject>git commit -m "Rolled back eggs.py to 009b7c0"
[master d41e595] Rolled back eggs.py to 009b7c0
 1 file changed, 47 deletions(-)

C:\Users\Al\myproject>git log --oneline
d41e595 (HEAD -> master) Rolled back eggs.py to 009b7c0
895d220 Adding email support to cheese().
df617da Renaming bacon() to cheese().
ef1e4bb Refactoring bacon().
ac27c9e Adding bacon() function.
009b7c0 Adding better documentation to spam().
0657588 Creating spam() function.
d811971 Initial add.
```

The *eggs.py* file has been rolled back, and the rest of the repo remains the same.

Rewriting the Commit History

If you've accidentally committed a file that contains sensitive information, such as passwords, API keys, or credit card numbers, it's not enough to edit that information out and make a new commit. Anyone with access to the repo, either on your computer or cloned remotely, could roll back to the commit that includes this info.

Actually removing this information from your repo so it's unrecoverable is tricky but possible. The exact steps are beyond the scope of this book, but you can use either the git filter-branch command or, preferably, the BFG Repo-Cleaner tool. You can read about both at *https://help.github .com/en/articles/removing-sensitive-data-from-a-repository*.

The easiest preventative measure for this problem is to have a *secrets.txt, confidential.py*, or similarly named file where you place sensitive, private information, and add it to *.gitignore* so you'll never accidentally commit it to the repo. Your program can read this file for the sensitive info instead of having the sensitive info directly in its source code.

GitHub and the git push Command

Although Git repos can exist entirely on your computer, many free websites can host clones of your repo online, letting others easily download and contribute to your projects. The largest of these sites is GitHub. If you keep a clone of your project online, others can add to your code, even if the computer from which you develop is turned off. The clone also acts as an effective backup.

NOTE *Although the terms can cause confusion,* Git *is version control software that maintains a repo and includes the* git *command, whereas* GitHub *is a website that hosts Git repos online.*

Go to *https://github.com* and sign up for a free account. From the GitHub home page or your profile page's Repositories tab, click the **New** button to start a new project. Enter **wizcoin** for the repository name and the same project description that we gave Cookiecutter in "Using Cookiecutter to Create New Python Projects" on page 200, as shown in Figure 12-6. Mark the repo as **Public** and deselect the **Initialize this repository with a README** checkbox, because we'll import an existing repository. Then click **Create repository**. These steps are effectively like running git init on the GitHub website.

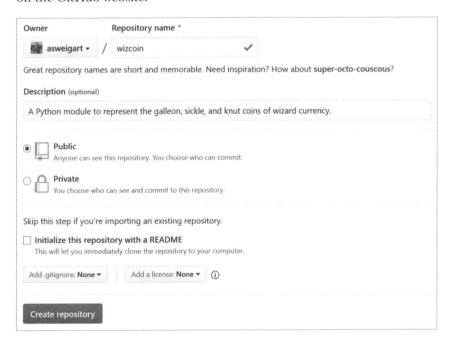

Figure 12-6: Creating a new repo on GitHub

You'll find the web page for your repos at *https://github.com/<username>/ <repo_name>*. In my case, my wizcoin repo is hosted at *https://github.com/ asweigart/wizcoin*.

Pushing an Existing Repository to GitHub

To push an existing repository from the command line, enter the following:

```
C:\Users\Al\wizcoin>git remote add origin https://github.com/<github_
username>/wizcoin.git
C:\Users\Al\wizcoin>git push -u origin master
Username for 'https://github.com': <github_username>
Password for 'https://<github_username>@github.com': <github_password>
Counting objects: 3, done.
Writing objects: 100% (3/3), 213 bytes | 106.00 KiB/s, done.
Total 3 (delta 0), reused 0 (delta 0)
To https://github.com/<your github>/wizcoin.git
 * [new branch]      master -> master
Branch 'master' set up to track remote branch 'master' from 'origin'.
```

The git remote add origin https://github.com/<github_username>/wizcoin.
git command adds GitHub as a remote repo corresponding to your local
repo. You then push any commits you've made on your local repo to the
remote repo using the git push -u origin master command. After this first
push, you can push all future commits from your local repo by simply running git push. Pushing your commits to GitHub after every commit is a
good idea to ensure the remote repo on GitHub is up to date with your
local repo, but it's not strictly necessary.

When you reload the repo's web page on GitHub, you should see the
files and commits displayed on the site. There's a lot more to learn about
GitHub, including how you can accept other people's contributions to your
repos through *pull requests*. These, along with GitHub's other advanced features, are beyond the scope of this book.

Cloning a Repo from an Existing GitHub Repo

It's also possible to do the opposite: create a new repo on GitHub and clone
it to your computer. Create a new repo on the GitHub website, but this time,
select the **Initialize this repository with a README** checkbox.

To clone this repo to your local computer, go to the repo's page on
GitHub and click the **Clone** or **download** button to open a window whose
URL should look something like *https://github.com/<github_username>/wizcoin.
git*. Use your repo's URL with the git clone command to download it to your
computer:

```
C:\Users\Al>git clone https://github.com/<github_username>/wizcoin.git
Cloning into 'wizcoin'...
remote: Enumerating objects: 5, done.
remote: Counting objects: 100% (5/5), done.
remote: Compressing objects: 100% (3/3), done.
remote: Total 5 (delta 0), reused 5 (delta 0), pack-reused 0
Unpacking objects: 100% (5/5), done.
```

You can now commit and push changes using this Git repo just as you
would if you had run git init to create the repo.

The git clone command is also useful in case your local repo gets into a state that you don't know how to undo. Although it's less than ideal, you can always save a copy of the files in your working directory, delete the local repo, and use git clone to re-create the repo. This scenario happens so often, even to experienced software developers, that it's the basis of the joke at *https://xkcd.com/1597/*.

Summary

Version control systems are lifesavers for programmers. Committing snapshots of your code makes it easy to review your progress and, in certain cases, roll back changes you don't need. Learning the basics of a version control system like Git certainly saves you time in the long run.

Python projects typically have several standard files and folders, and the cookiecutter module helps you create the starting boilerplate for many of these files. These files make up the first files you commit to your local Git repo. We call the folder containing all of this content the *working directory* or *project folder.*

Git tracks the files in your working directory, all of which can exist in one of three states: committed (also called unmodified or clean), modified, or staged. The Git command line tool has several commands, such as git status or git log, that let you view this information, but you can also install several third-party GUI Git tools.

The git init command creates a new, empty repo on your local computer. The git clone command copies a repo from a remote server, such as the popular GitHub website. Either way, once you have a repo, you can use git add and git commit to commit changes to your repo, and use git push to push these commits to a remote GitHub repo. Several commands were also described in this chapter to undo the commits you've made. Performing an undo allows you to roll back to an earlier version of your files.

Git is a extensive tool with many features, and this chapter covers only the basics of the version control system. Many resources are available to you to learn more about Git's advanced features. I recommend two free books that you can find online: *Pro Git* by Scott Charcon at *https://git-scm.com/book/en/v2* and *Version Control by Example* by Eric Sink at *https://ericsink.com/vcbe/index.html.*

13

MEASURING PERFORMANCE AND BIG O ALGORITHM ANALYSIS

For most small programs, performance isn't all that important. We might spend an hour writing a script to automate a task that needs only a few seconds to run. Even if it takes longer, the program will probably finish by the time we've returned to our desk with a cup of coffee.

Sometimes it's prudent to invest time in learning how to make a script faster. But we can't know if our changes improve a program's speed until we know how to measure program speed. This is where Python's timeit and cProfile modules come in. These modules not only measure how fast code runs, but also create a *profile* of which parts of the code are already fast and which parts we could improve.

In addition to measuring program speed, in this chapter you'll also learn how to measure the theoretical increases in runtime as the data for your program grows. In computer science, we call this *big O notation*. Software developers without traditional computer science backgrounds might sometimes feel they have gaps in their knowledge. But although a computer science education is fulfilling, it's not always directly relevant to software development. I joke (but only half so) that big O notation makes up about 80 percent of the usefulness of my degree. This chapter provides an introduction to this practical topic.

The timeit Module

"Premature optimization is the root of all evil" is a common saying in software development. (It's often attributed to computer scientist Donald Knuth, who attributes it to computer scientist Tony Hoare. Tony Hoare, in turn, attributes it to Donald Knuth.) *Premature optimization*, or optimizing before knowing what needs to be optimized, often manifests itself when programmers try to use clever tricks to save memory or write faster code. For example, one of these tricks is using the *XOR algorithm* to swap two integer values without using a third, temporary variable:

```
>>> a, b = 42, 101  # Set up the two variables.
>>> print(a, b)
42 101
>>> # A series of ^ XOR operations will end up swapping their values:
>>> a = a ^ b
>>> b = a ^ b
>>> a = a ^ b
>>> print(a, b)  # The values are now swapped.
101 42
```

Unless you're unfamiliar with the XOR algorithm (which uses the ^ bit-wise operator), this code looks cryptic. The problem with using clever programming tricks is that they can produce complicated, unreadable code. Recall that one of the Zen of Python tenets is *readability counts.*

Even worse, your clever trick might turn out not to be so clever. You can't just assume a crafty trick is faster or that the old code it's replacing was even all that slow to begin with. The only way to find out is by measuring and comparing the *runtime*: the amount of time it takes to run a program or piece of code. Keep in mind that increasing the runtime means the program is slowing down: the program is taking more time to do the same amount of work. (We also sometimes use the term *runtime* to refer to the period during which the program is running. *This error happened at runtime* means the error happened while the program was running as opposed to when the program was being compiled into bytecode.)

The Python standard library's timeit module can measure the runtime speed of a small snippet of code by running it thousands or millions of times, letting you determine an average runtime. The timeit module also temporarily disables the automatic garbage collector to ensure more consistent runtimes. If you want to test multiple lines, you can pass a multiline code string or separate the code lines using semicolons:

```
>>> import timeit
>>> timeit.timeit('a, b = 42, 101; a = a ^ b; b = a ^ b; a = a ^ b')
0.1307766629999998
>>> timeit.timeit("""a, b = 42, 101
```

```
...     a = a ^ b
...     b = a ^ b
...     a = a ^ b""")
0.13515726800000039
```

On my computer, the XOR algorithm takes roughly one-tenth of a second to run this code. Is this fast? Let's compare it to some integer swapping code that uses a third temporary variable:

```
>>> import timeit
>>> timeit.timeit('a, b = 42, 101; temp = a; a = b; b = temp')
0.027540389999998638
```

That's a surprise! Not only is using a third temporary variable more readable, but it's also twice as fast! The *clever* XOR trick might save a few bytes of memory but at the expense of speed and code readability. Sacrificing code readability to reduce a few bytes of memory usage or nanoseconds of runtime isn't worthwhile.

Better still, you can swap two variables using the *multiple assignment trick*, also called *iterable unpacking*, which also runs in a small amount of time:

```
>>> timeit.timeit('a, b = 42, 101; a, b = b, a')
0.024489236000007963
```

Not only is this the most readable code, it's also the quickest. We know this not because we assumed it, but because we objectively measured it.

The timeit.timeit() function can also take a second string argument of *setup* code. The setup code runs only once before running the first string's code. You can also change the default number of trials by passing an integer for the number keyword argument. For example, the following test measures how quickly Python's random module can generate 10,000,000 random numbers from 1 to 100. (On my machine, it takes about 10 seconds.)

```
>>> timeit.timeit('random.randint(1, 100)', 'import random', number=10000000)
10.020913950999784
```

By default, the code in the string you pass to timeit.timeit() won't be able to access the variables and the functions in the rest of the program:

```
>>> import timeit
>>> spam = 'hello'  # We define the spam variable.
>>> timeit.timeit('print(spam)', number=1)  # We measure printing spam.
Traceback (most recent call last):
  File "<stdin>", line 1, in <module>
  File "C:\Users\Al\AppData\Local\Programs\Python\Python37\lib\timeit.py",
line 232, in timeit
    return Timer(stmt, setup, timer, globals).timeit(number)
  File "C:\Users\Al\AppData\Local\Programs\Python\Python37\lib\timeit.py",
line 176, in timeit
    timing = self.inner(it, self.timer)
  File "<timeit-src>", line 6, in inner
NameError: name 'spam' is not defined
```

To fix this, pass the function the return value of globals() for the globals keyword argument:

```
>>> timeit.timeit('print(spam)', number=1, globals=globals())
hello
0.000994909999462834
```

A good rule for writing your code is to first make it work and then make it fast. Only once you have a working program should you focus on making it more efficient.

The cProfile Profiler

Although the timeit module is useful for measuring small code snippets, the cProfile module is more effective for analyzing entire functions or programs. *Profiling* analyzes your program's speed, memory usage, and other aspects systematically. The cProfile module is Python's *profiler*, or software that can measure a program's runtime as well as build a profile of runtimes for the program's individual function calls. This information provides substantially more granular measurements of your code.

To use the cProfile profiler, pass a string of the code you want to measure to cProfile.run(). Let's look at how cProfiler measures and reports the execution of a short function that sums all the numbers from 1 to 1,000,000:

```
import time, cProfile
def addUpNumbers():
    total = 0
    for i in range(1, 1000001):
        total += i

cProfile.run('addUpNumbers()')
```

When you run this program, the output will look something like this:

```
         4 function calls in 0.064 seconds

   Ordered by: standard name

   ncalls  tottime  percall  cumtime  percall filename:lineno(function)
        1    0.000    0.000    0.064    0.064 <string>:1(<module>)
        1    0.064    0.064    0.064    0.064 test1.py:2(addUpNumbers)
        1    0.000    0.000    0.064    0.064 {built-in method builtins.exec}
        1    0.000    0.000    0.000    0.000 {method 'disable' of '_lsprof.Profiler' objects}
```

Each line represents a different function and the amount of time spent in that function. The columns in cProfile.run()'s output are:

ncalls The number of calls made to the function

tottime The total time spent in the function, excluding time in subfunctions

> *percall*   The total time divided by the number of calls

> *cumtime*   The cumulative time spent in the function and all subfunctions

> *percall*   The cumulative time divided by the number of calls

> *filename:lineno(function)*   The file the function is in and at which line number

For example, download the *rsaCipher.py* and *al_sweigart_pubkey.txt* files from *https://nostarch.com/crackingcodes/*. This RSA Cipher program was featured in *Cracking Codes with Python* (No Starch Press, 2018). Enter the following into the interactive shell to profile the encryptAndWriteToFile() function as it encrypts a 300,000-character message created by the 'abc' * 100000 expression:

```
>>> import cProfile, rsaCipher
>>> cProfile.run("rsaCipher.encryptAndWriteToFile('encrypted_file.txt', 'al_sweigart_pubkey.
txt', 'abc'*100000)")
        11749 function calls in 28.900 seconds

   Ordered by: standard name

   ncalls  tottime  percall  cumtime  percall filename:lineno(function)
        1    0.001    0.001   28.900   28.900 <string>:1(<module>)
        2    0.000    0.000    0.000    0.000 _bootlocale.py:11(getpreferredencoding)
--snip--
        1    0.017    0.017   28.900   28.900 rsaCipher.py:104(encryptAndWriteToFile)
        1    0.248    0.248    0.249    0.249 rsaCipher.py:36(getBlocksFromText)
        1    0.006    0.006   28.873   28.873 rsaCipher.py:70(encryptMessage)
        1    0.000    0.000    0.000    0.000 rsaCipher.py:94(readKeyFile)
--snip--
     2347    0.000    0.000    0.000    0.000 {built-in method builtins.len}
     2344    0.000    0.000    0.000    0.000 {built-in method builtins.min}
     2344   28.617    0.012   28.617    0.012 {built-in method builtins.pow}
        2    0.001    0.000    0.001    0.000 {built-in method io.open}
     4688    0.001    0.000    0.001    0.000 {method 'append' of 'list' objects}
--snip--
```

You can see that the code passed to cProfile.run() took 28.9 seconds to complete. Pay attention to the functions with the highest total times; in this case, Python's built-in pow() function takes up 28.617 seconds. That's nearly the entire code's runtime! We can't change this code (it's part of Python), but perhaps we could change our code to rely on it less.

This isn't possible in this case, because *rsaCipher.py* is already fairly optimized. Even so, profiling this code has provided us insight that pow() is the main bottleneck. So there's little sense in trying to improve, say, the readKeyFile() function (which takes so little time to run that cProfile reports its runtime as 0).

This idea is captured by *Amdahl's Law*, a formula that calculates how much the overall program speeds up given an improvement to one of its components. The formula is *speed-up of whole task = 1 / ((1 − p) + (p / s))* where *s* is the speed-up made to a component and *p* is the portion of that component of the overall program. So if you double the speed of a component that makes up 90 percent of the program's total runtime, you'll get $1 / ((1 - 0.9) + (0.9 / 2)) = 1.818$, or an 82 percent speed-up of the overall program. This is better than, say, tripling the speed of a component that only makes up 25 percent of the total runtime, which would only be a $1 / ((1 - 0.25) + (0.25 / 2)) = 1.143$, or 14 percent overall speed-up. You don't need to memorize the formula. Just remember that doubling the speed of your code's slow or lengthy parts is more productive than doubling the speed of an already quick or short part. This is common sense: a 10 percent price discount on an expensive house is better than a 10 percent discount on a cheap pair of shoes.

Big O Algorithm Analysis

Big O is a form of algorithm analysis that describes how code will scale. It classifies the code into one of several orders that describes, in general terms, how much longer the code's runtime will take as the amount of work it has to do increases. Python developer Ned Batchelder describes big O as an analysis of "how code slows as data grows," which is also the title of his informative PyCon 2018 talk, which is available at *https://youtu.be/duvZ-2UK0fc/*.

Let's consider the following scenario. Say you have a certain amount of work that takes an hour to complete. If the workload doubles, how long would it take then? You might be tempted to think it takes twice as long, but actually, the answer depends on the kind of work that's being done.

If it takes an hour to read a short book, it will take more or less two hours to read two short books. But if you can alphabetize 500 books in an hour, alphabetizing 1,000 books will most likely take longer than two hours, because you have to find the correct place for each book in a much larger collection of books. On the other hand, if you're just checking whether or not a bookshelf is empty, it doesn't matter if there are 0, 10, or 1,000 books on the shelf. One glance and you'll immediately know the answer. The runtime remains roughly constant no matter how many books there are. Although some people might be faster or slower at reading or alphabetizing books, these general trends remain the same.

The big O of the algorithm describes these trends. An algorithm can run on a fast or slow computer, but we can still use big O to describe how well an algorithm performs in general, regardless of the actual hardware executing the algorithm. Big O doesn't use specific units, such as seconds or CPU cycles, to describe an algorithm's runtime, because these would vary between different computers or programming languages.

Big O Orders

Big O notation commonly defines the following orders. These range from the *lower* orders, which describe code that, as the amount of data grows, slows down the least, to the *higher* orders, which describe code that slows down the most:

1. O(1), Constant Time (the lowest order)
2. O(log n), Logarithmic Time
3. O(n), Linear Time
4. O(n log n), N-Log-N Time
5. O(n^2), Polynomial Time
6. O(2^n), Exponential Time
7. O(n!), Factorial Time (the highest order)

Notice that big O uses the following notation: a capital O, followed by a pair of parentheses containing a description of the order. The capital O represents *order* or *on the order of*. The n represents the size of the input data the code will work on. We pronounce O(n) as "big oh of n" or "big oh n."

You don't need to understand the precise mathematic meanings of words like *logarithmic* or *polynomial* to use big O notation. I'll describe each of these orders in detail in the next section, but here's an oversimplified explanation of them:

- O(1) and O(log n) algorithms are fast.
- O(n) and O(n log n) algorithms aren't bad.
- O(n^2), O(2^n), and O(n!) algorithms are slow.

Certainly, you could find counterexamples, but these descriptions are good rules in general. There are more big O orders than the ones listed here, but these are the most common. Let's look at the kinds of tasks that each of these orders describes.

A Bookshelf Metaphor for Big O Orders

In the following big O order examples, I'll continue using the bookshelf metaphor. The *n* refers to the number of books on the bookshelf, and the big O ordering describes how the various tasks take longer as the number of books increases.

O(1), Constant Time

Finding out "Is the bookshelf empty?" is a constant time operation. It doesn't matter how many books are on the shelf; one glance tells us whether or not the bookshelf is empty. The number of books can vary, but the runtime remains constant, because as soon as we see one book on the shelf, we can stop looking. The *n* value is irrelevant to the speed of the task, which is why there is no *n* in O(1). You might also see constant time written as O(c).

O(log n), Logarithmic

Logarithms are the inverse of exponentiation: the exponent 2^4, or $2 \times 2 \times 2 \times 2$, equals 16, whereas the logarithm $\log_2(16)$ (pronounced "log base 2 of 16") equals 4. In programming, we often assume base 2 to be the logarithm base, which is why we write O(log n) instead of O($\log_2$ n).

Searching for a book on an alphabetized bookshelf is a logarithmic time operation. To find one book, you can check the book in the middle of the shelf. If that is the book you're searching for, you're done. Otherwise, you can determine whether the book you're searching for comes before or after this middle book. By doing so, you've effectively reduced the range of books you need to search in half. You can repeat this process again, checking the middle book in the half that you expect to find it. We call this the *binary search algorithm*, and there's an example of it in "Big O Analysis Examples" later in this chapter.

The number of times you can split a set of *n* books in half is $\log_2$ n. On a shelf of 16 books, it will take at most four steps to find the right one. Because each step reduces the number of books you need to search by one half, a bookshelf with double the number of books takes just one more step to search. If there were 4.2 billion books on the alphabetized bookshelf, it would still only take 32 steps to find a particular book.

Log *n* algorithms usually involve a *divide and conquer* step, which selects half of the *n* input to work on and then another half from that half, and so on. Log *n* operations scale nicely: the workload *n* can double in size, but the runtime increases by only one step.

O(n), Linear Time

Reading all the books on a bookshelf is a linear time operation. If the books are roughly the same length and you double the number of books on the shelf, it will take roughly double the amount of time to read all the books. The runtime increases *in proportion* to the number of books *n*.

O(n log n), N-Log-N Time

Sorting a set of books into alphabetical order is an n-log-n time operation. This order is the runtime of O(n) and O(log n) multiplied together. You can think of a O(n log n) task as a O(log n) task that must be performed *n* times. Here's a casual description of why.

Start with a stack of books to alphabetize and an empty bookshelf. Follow the steps for a binary search algorithm, as detailed in "O(log n), Logarithmic" on page 232, to find where a single book belongs on the shelf. This is an O(log n) operation. With *n* books to alphabetize, and each book taking log *n* steps to alphabetize, it takes *n* × log *n*, or *n* log *n*, steps to alphabetize the entire set of books. Given twice as many books, it takes a bit more than twice as long to alphabetize all of them, so *n* log *n* algorithms scale fairly well.

In fact, all of the efficient general sorting algorithms are O(n log n): merge sort, quicksort, heapsort, and Timsort. (Timsort, invented by Tim Peters, is the algorithm that Python's sort() method uses.)

$O(n^2)$, Polynomial Time

Checking for duplicate books on an unsorted bookshelf is a polynomial time operation. If there are 100 books, you could start with the first book and compare it with the 99 other books to see whether they're the same. Then you take the second book and check whether it's the same as any of the 99 other books. Checking for a duplicate of a single book is 99 steps (we'll round this up to 100, which is our n in this example). We have to do this 100 times, once for each book. So the number of steps to check for any duplicate books on the bookshelf is roughly $n \times n$, or n^2. (This approximation to n^2 still holds even if we were clever enough not to repeat comparisons.)

The runtime increases by the increase in books squared. Checking 100 books for duplicates is 100×100, or 10,000 steps. But checking twice that amount, 200 books, is 200×200, or 40,000 steps: four times as much work.

In my experience writing code in the real world, I've found the most common use of big O analysis is to avoid accidentally writing an $O(n^2)$ algorithm when an O(n log n) or O(n) algorithm exists. The $O(n^2)$ order is when algorithms dramatically slow down, so recognizing your code as $O(n^2)$ or higher should give you pause. Perhaps there's a different algorithm that can solve the problem faster. In these cases, taking a data structure and algorithms (DSA) course, whether at a university or online, can be helpful.

We also call $O(n^2)$ *quadratic time*. Algorithms could have $O(n^3)$, or *cubic time*, which is slower than $O(n^2)$; $O(n^4)$, or *quartic time*, which is slower than $O(n^3)$; or other polynomial time complexities.

$O(2^n)$, Exponential Time

Taking photos of the shelf with every possible combination of books on it is an exponential time operation. Think of it this way: each book on the shelf can either in be included in the photo or not included. Figure 13-1 shows every combination where n is 1, 2, or 3. If n is 1, there are two possible photos: with the book and without. If n is 2, there are four possible photos: both books on the shelf, both books off, the first on and second off, and the second on and first off. When you add a third book, you've once again doubled the amount of work you have to do: you need to do every subset of two books that includes the third book (four photos) and every subset of two books that excludes the third book (another four photos, for 2^3 or eight photos). Each additional book doubles the workload. For n books, the number of photos you need to take (that is, the amount of work you need to do) is 2^n.

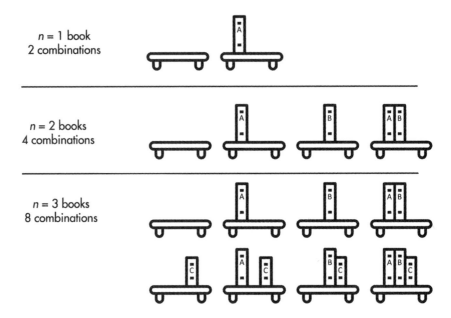

n = 1 book
2 combinations

n = 2 books
4 combinations

n = 3 books
8 combinations

Figure 13-1: Every combination of books on a bookshelf for one, two, or three books

The runtime for exponential tasks increases very quickly. Six books require 2^6 or 32 photos, but 32 books would include 2^{32} or more than 4.2 billion photos. $O(2^n)$, $O(3^n)$, $O(4^n)$, and so on are different orders ut all have exponential time complexity.

O(n!), Factorial Time

Taking a photo of the books on the shelf in every conceivable order is a factorial time operation. We call every possible order the *permutation* of n books. This results in $n!$, or n *factorial*, orderings. The *factorial* of a number is the multiplication product of all positive integers up to the number. For example, $3!$ is $3 \times 2 \times 1$, or 6. Figure 13-2 shows every possible permutation of three books.

Figure 13-2: All 3! (that is, 6) permutations of three books on a bookshelf

To calculate this yourself, think about how you would come up with every permutation of n books. You have n possible choices for the first book; then $n - 1$ possible choices for the second book (that is, every book except the one you picked for the first book); then $n - 2$ possible choices for the third book; and so on. With 6 books, 6! results in $6 \times 5 \times 4 \times 3 \times 2 \times 1$, or 720 photos. Adding just one more book makes this 7!, or 5,040 photos needed. Even for small *n* values, factorial time algorithms quickly become impossible to complete in a reasonable amount of time. If you had 20 books and could arrange them and take a photo every second, it would still take longer than the universe has existed to get through every possible permutation.

One well-known O(n!) problem is the traveling salesperson conundrum. A salesperson must visit *n* cities and wants to calculate the distance travelled for all *n!* possible orders in which they could visit them. From these calculations, they could determine the order that involves the shortest travel distance. In a region with many cities, this task proves impossible to complete in a timely way. Fortunately, optimized algorithms can find a short (but not guaranteed to be the shortest) route much faster than O(n!).

Big O Measures the Worst-Case Scenario

Big O specifically measures the *worst-case scenario* for any task. For example, finding a particular book on an unorganized bookshelf requires you to start from one end and scan the books until you find it. You might get lucky, and the book you're looking for might be the first book you check. But you might be unlucky; it could be the last book you check or not on the bookshelf at all. So in a best-case scenario, it wouldn't matter if there were billions of books you had to search through, because you'll immediately find the one you're looking for. But that optimism isn't useful for algorithm analysis. Big O describes what happens in the unlucky case: if you have *n* books on the shelf, you'll have to check all *n* books. In this example, the runtime increases at the same rate as the number of books.

Some programmers also use *big Omega notation*, which describes the best-case scenario for an algorithm. For example, a $\Omega(n)$ algorithm performs at linear efficiency at its best. In the worst case, it might perform slower. Some algorithms encounter especially lucky cases where no work has to be done, such as finding driving directions to a destination when you're already at the destination.

Big Theta notation describes algorithms that have the same best- and worst-case order. For example, $\Theta(n)$ describes an algorithm that has linear efficiency at best *and* at worst, which is to say, it's an O(n) and $\Omega(n)$ algorithm. These notations aren't used in software engineering as often as big O, but you should still be aware of their existence.

It isn't uncommon for people to talk about the "big O of the average case" when they mean big Theta, or "big O of the best case" when they mean big Omega. This is an oxymoron; big O specifically refers to an algorithm's worst-case runtime. But even though their wording is technically incorrect, you can understand their meaning irregardless.

MORE THAN ENOUGH MATH TO DO BIG O

If your algebra is rusty, here's more than enough math to do big O analysis:

Multiplication   Repeated addition. $2 \times 4 = 8$, just like $2 + 2 + 2 + 2 = 8$. With variables, $n + n + n$ is $3 \times n$.

Multiplication notation   Algebra notation often omits the $\times$ sign, so $2 \times n$ is written as $2n$. With numbers, 2×3 is written as $2(3)$ or simply 6.

The multiplicative identity property   Multiplying a number by 1 results in that number: $5 \times 1 = 5$ and $42 \times 1 = 42$. More generally, $n \times 1 = n$.

The distributive property of multiplication $2 \times (3 + 4) = (2 \times 3) + (2 \times 4)$. Both sides of the equation equal 14. More generally, $a(b + c) = ab + ac$.

Exponentiation   Repeated multiplication. $2^4 = 16$ (pronounced "2 raised to the 4th power is 16"), just like $2 \times 2 \times 2 \times 2 = 16$. Here, 2 is the *base* and 4 is the *exponent*. With variables, $n \times n \times n \times n$ is n^4. In Python, we use the ** operator: `2 ** 4` evaluates to 16.

The 1st power evaluates to the base $2^1 = 2$ and $9999^1 = 9999$. More generally, $n^1 = n$.

The 0th power always evaluates to 1 $2^0 = 1$ and $9999^0 = 1$. More generally, $n^0 = 1$.

Coefficients   Multiplicative factors. In $3n^2 + 4n + 5$, the coefficients are 3, 4, and 5. You can see that 5 is a coefficient because 5 can be rewritten as $5(1)$ and then rewritten as $5n^0$.

Logarithms   The inverse of exponentiation. Because $2^4 = 16$, we know that $\log_2(16) = 4$. We pronounce this "the log base 2 of 16 is 4." In Python, we use the `math.log()` function: `math.log(16, 2)` evaluates to `4.0`.

Calculating big O often involves simplifying equations by combining like terms. A *term* is some combination of numbers and variables multiplied together: in $3n^2 + 4n + 5$, the terms are $3n^2$, $4n$, and 5. *Like terms* have the same variables raised to the same exponent. In the expression $3n^2 + 4n + 6n + 5$, the terms $4n$ and $6n$ are like terms. We could simplify and rewrite this as $3n^2 + 10n + 5$.

Keep in mind that because $n \times 1 = n$, an expression like $3n^2 + 5n + 4$ can be thought of as $3n^2 + 5n + 4(1)$. The terms in this expression match with the big O orders O(n^2), O(n), and O(1). This will come up later when we're dropping coefficients for our big O calculations.

These math rule reminders might come in handy when you're first learning how to figure out the big O of a piece of code. But by the time you finish "Analyzing Big O at a Glance" later in this chapter, you probably won't need them anymore. Big O is a simple concept and can be useful even if you don't strictly follow mathematical rules.

Determining the Big O Order of Your Code

To determine the big O order for a piece of code, we must do four tasks: identify what the *n* is, count the steps in the code, drop the lower orders, and drop the coefficients.

For example, let's find the big O of the following readingList() function:

```
def readingList(books):
    print('Here are the books I will read:')
    numberOfBooks = 0
    for book in books:
        print(book)
        numberOfBooks += 1
    print(numberOfBooks, 'books total.')
```

Recall that the *n* represents the size of the input data that the code works on. In functions, the *n* is almost always based on a parameter. The readingList() function's only parameter is books, so the size of books seems like a good candidate for *n*, because the larger books is, the longer the function takes to run.

Next, let's count the steps in this code. What counts as a *step* is somewhat vague, but a line of code is a good rule to follow. Loops will have as many steps as the number of iterations multiplied by the lines of code in the loop. To see what I mean, here are the counts for the code inside the readingList() function:

```
def readingList(books):
    print('Here are the books I will read:')    # 1 step
    numberOfBooks = 0                           # 1 step
    for book in books:                          # n * steps in the loop
        print(book)                             # 1 step
        numberOfBooks += 1                      # 1 step
    print(numberOfBooks, 'books total.')        # 1 step
```

We'll treat each line of code as one step except for the for loop. This line executes once for each item in books, and because the size of books is our *n*, we say that this executes *n* steps. Not only that, but it executes all the steps inside the loop *n* times. Because there are two steps inside the loop, the total is $2 \times n$ steps. We could describe our steps like this:

```
def readingList(books):
    print('Here are the books I will read:')    # 1 step
    numberOfBooks = 0                           # 1 step
    for book in books:                          # n * 2 steps
        print(book)                             # (already counted)
        numberOfBooks += 1                      # (already counted)
    print(numberOfBooks, 'books total.')        # 1 step
```

Now when we compute the total number of steps, we get $1 + 1 + (n \times 2) + 1$. We can rewrite this expression more simply as $2n + 3$.

Big O doesn't intend to describe specifics; it's a general indicator. Therefore, we drop the lower orders from our count. The orders in $2n + 3$ are linear ($2n$) and constant (3). If we keep only the largest of these orders, we're left with $2n$.

Next, we drop the coefficients from the order. In $2n$, the coefficient is 2. After dropping it, we're left with n. This gives us the final big O of the readingList() function: O(n), or linear time complexity.

This order should make sense if you think about it. There are several steps in our function, but in general, if the books list increases tenfold in size, the runtime increases about tenfold as well. Increasing books from 10 books to 100 books moves the algorithm from $1 + 1 + (2 \times 10) + 1$, or 23 steps, to $1 + 1 + (2 \times 100) + 1$, or 203 steps. The number 203 is roughly 10 times 23, so the runtime increases proportionally with the increase to n.

Why Lower Orders and Coefficients Don't Matter

We drop the lower orders from our step count because they become less significant as n grows in size. If we increased the books list in the previous readingList() function from 10 to 10,000,000,000 (10 billion), the number of steps would increase from 23 to 20,000,000,003. With a large enough n, those extra three steps matter very little.

When the amount of data increases, a large coefficient for a smaller order won't make a difference compared to the higher orders. At a certain size n, the higher orders will always be slower than the lower orders. For example, let's say we have quadraticExample(), which is O(n^2) and has $3n^2$ steps. We also have linearExample(), which is O(n) and has $1,000n$ steps. It doesn't matter that the 1,000 coefficient is larger than the 3 coefficient; as n increases, eventually an O(n^2) quadratic operation will become slower than an O(n) linear operation. The actual code doesn't matter, but we can think of it as something like this:

```
def quadraticExample(someData):   # n is the size of someData
    for i in someData:            # n steps
        for j in someData:        # n steps
            print('Something')    # 1 step
            print('Something')    # 1 step
            print('Something')    # 1 step

def linearExample(someData):      # n is the size of someData
    for i in someData:            # n steps
        for k in range(1000):     # 1 * 1000 steps
            print('Something')    # (Already counted)
```

The linearExample() function has a large coefficient (1,000) compared to the coefficient (3) of quadraticExample(). If the size of the input n is 10, the O(n^2) function appears faster with only 300 steps compared to the O(n) function with 10,000 steps.

But big O notation is chiefly concerned with the algorithm's performance as the workload scales up. When n reaches the size of 334 or greater, the quadraticExample() function will always be slower than the linearExample() function. Even if linearExample() were 1,000,000n steps, the quadraticExample() function would still become slower once n reached 333,334. At some point, an O(n²) operation always becomes slower than an O(n) or lower operation. To see how, look at the big O graph shown in Figure 13-3. This graph features all the major big O notation orders. The x-axis is n, the size of the data, and the y-axis is the runtime needed to carry out the operation.

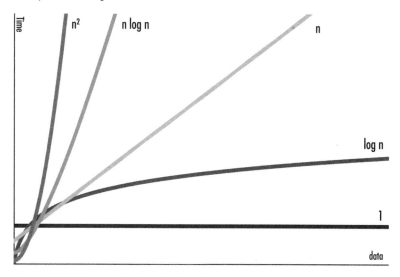

Figure 13-3: The graph of the big O orders

As you can see, the runtime of the higher orders grows at a faster rate than that of the lower orders. Although the lower orders could have large coefficients that make them temporarily larger than the higher orders, the higher orders eventually outpace them.

Big O Analysis Examples

Let's determine the big O orders of some example functions. In these examples, we'll use a parameter named books that is a list of strings of book titles.

The countBookPoints() function calculates a score based on the number of books in the books list. Most books are worth one point, and books by a certain author are worth two points:

```
def countBookPoints(books):
    points = 0           # 1 step
    for book in books:   # n * steps in the loop
        points += 1      # 1 step

    for book in books:                    # n * steps in the loop
        if 'by Al Sweigart' in book:      # 1 step
            points += 1                   # 1 step
    return points                         # 1 step
```

The number of steps comes to $1 + (n \times 1) + (n \times 2) + 1$, which becomes $3n + 2$ after combining the like terms. Once we drop the lower orders and coefficients, this becomes O(n), or linear complexity, no matter if we loop through books once, twice, or a billion times.

So far, all examples that used a single loop have had linear complexity, but notice that these loops iterated n times. As you'll see in the next example, a loop in your code alone doesn't imply linear complexity, although a loop that iterates over your data does.

This iLoveBooks() function prints "I LOVE BOOKS!!!" and "BOOKS ARE GREAT!!!" 10 times:

```
def iLoveBooks(books):
    for i in range(10):                   # 10 * steps in the loop
        print('I LOVE BOOKS!!!')          # 1 step
        print('BOOKS ARE GREAT!!!')       # 1 step
```

This function has a for loop, but it doesn't loop over the books list, and it performs 20 steps no matter what the size of books is. We can rewrite this as 20(1). After dropping the 20 coefficient, we are left with O(1), or constant time complexity. This makes sense; the function takes the same amount of time to run, no matter what n, the size of the books list, is.

Next, we have a cheerForFavoriteBook() function that searches through the books list to look for a favorite book:

```
def cheerForFavoriteBook(books, favorite):
    for book in books:                            # n * steps in the loop
        print(book)                               # 1 step
        if book == favorite:                      # 1 step
            for i in range(100):                  # 100 * steps in the loop
                print('THIS IS A GREAT BOOK!!!')  # 1 step
```

The for book loop iterates over the books list, which requires n steps multiplied by the steps inside the loop. This loop includes a nested for i loop, which iterates 100 times. This means the for book loop runs $102 \times n$, or $102n$ steps. After dropping the coefficient, we find that cheerForFavoriteBook() is still just an O(n) linear operation. This 102 coefficient might seem rather large to just ignore, but consider this: if favorite never appears in the books list, this function would only run $1n$ steps. The impact of coefficients can vary so wildly that they aren't very meaningful.

Next, the findDuplicateBooks() function searches through the books list (a linear operation) once for each book (another linear operation):

```
def findDuplicateBooks(books):
    for i in range(books):  # n steps
        for j in range(i + 1, books):      # n steps
            if books[i] == books[j]:       # 1 step
                print('Duplicate:', books[i])  # 1 step
```

The for i loop iterates over the entire books list, performing the steps inside the loop n times. The for j loop also iterates over a portion of the books list, although because we drop coefficients, this also counts as a linear time operation. This means the for i loop performs $n \times n$ operations—that is, n^2. This makes findDuplicateBooks() an $O(n^2)$ polynomial time operation.

Nested loops alone don't imply a polynomial operation, but nested loops where both loops iterate n times do. These result in n^2 steps, implying an $O(n^2)$ operation.

Let's look at a challenging example. The binary search algorithm mentioned earlier works by searching the middle of a sorted list (we'll call it the haystack) for an item (we'll call it the needle). If we don't find the needle there, we'll proceed to search the previous or latter half of the haystack, depending on which half we expect to find the needle in. We'll repeat this process, searching smaller and smaller halves until either we find the needle or we conclude it isn't in the haystack. Note that binary search only works if the items in the haystack are in sorted order.

```
def binarySearch(needle, haystack):
    if not len(haystack):                       # 1 step
        return None                             # 1 step
    startIndex = 0                              # 1 step
    endIndex = len(haystack) - 1               # 1 step

    haystack.sort()                            # ??? steps

    while start <= end:  # ??? steps
        midIndex = (startIndex + endIndex) // 2 # 1 step
        if haystack[midIndex] == needle:        # 1 step
            # Found the needle.
            return midIndex                     # 1 step
        elif needle < haystack[midIndex]:       # 1 step
            # Search the previous half.
            endIndex = midIndex - 1             # 1 step
        elif needle > haystack[mid]:            # 1 step
            # Search the latter half.
            startIndex = midIndex + 1           # 1 step
```

Two of the lines in binarySearch() aren't easy to count. The haystack .sort() method call's big O order depends on the code inside Python's sort() method. This code isn't very easy to find, but you can look up its big O order on the internet to learn that it's $O(n \log n)$. (All general sorting functions are, at best, $O(n \log n)$.) We'll cover the big O order of several common Python functions and methods in "The Big O Order of Common Function Calls" later in this chapter.

The while loop isn't as straightforward to analyze as the for loops we've seen. We must understand the binary search algorithm to determine how many iterations this loop has. Before the loop, the startIndex and endIndex cover the entire range of haystack, and midIndex is set to the midpoint of this range. On each iteration of the while loop, one of two things happens. If haystack[midIndex] == needle, we know we've found the needle, and the function returns the index of the needle in haystack. If needle < haystack[midIndex] or needle > haystack[midIndex], the range covered by startIndex and endIndex is halved, either by adjusting startIndex or adjusting endIndex. The number of times we can divide any list of size *n* in half is $\log_2(n)$. (Unfortunately, this is simply a mathematical fact that you'd be expected to know.) Thus, the `while` loop has a big O order of O(log n).

But because the O(n log n) order of the `haystack.sort()` line is higher than O(log n), we drop the lower O(log n) order, and the big O order of the *entire* binarySearch() function becomes O(n log n). If we can guarantee that binarySearch() will only ever be called with a sorted list for haystack, we can remove the `haystack.sort()` line and make binarySearch() an O(log n) function. This technically improves the function's efficiency but doesn't make the overall program more efficient, because it just moves the required sorting work to some other part of the program. Most binary search implementations leave out the sorting step and therefore the binary search algorithm is said to have O(log n) logarithmic complexity.

The Big O Order of Common Function Calls

Your code's big O analysis must consider the big O order of any functions it calls. If you wrote the function, you can just analyze your own code. But to find the big O order of Python's built-in functions and methods, you'll have to consult lists like the following.

This list contains the big O orders of some common Python operations for sequence types, such as strings, tuples, and lists:

s[i] reading and s[i] = value assignment   O(1) operations.

s.append(value)   An O(1) operation.

s.insert(i, value)   An O(n) operation. Inserting values into a sequence (especially at the front) requires shifting all the items at indexes above i up by one place in the sequence.

s.remove(value)   An O(n) operation. Removing values from a sequence (especially at the front) requires shifting all the items at indexes above I down by one place in the sequence.

s.reverse()   An O(n) operation, because every item in the sequence must be rearranged.

s.sort()   An O(n log n) operation, because Python's sorting algorithm is O(n log n).

value in s   An O(n) operation, because every item must be checked.

for value in s: An O(n) operation.

len(s)   An O(1) operation, because Python keeps track of how many items are in a sequence so it doesn't need to recount them when passed to len().

This list contains the big O orders of some common Python operations for mapping types, such as dictionaries, sets, and frozensets:

m[key] reading and m[key] = value assignment   O(1) operations.

m.add(value)   An O(1) operation.

value in m   An O(1) operation for dictionaries, which is much faster than using *in* with sequences.

for key in m: An O(n) operation.

len(m)   An O(1) operation, because Python keeps track of how many items are in a mapping, so it doesn't need to recount them when passed to len().

Although a list generally has to search through its items from start to finish, dictionaries use the key to calculate the address, and the time needed to look up the key's value remains constant. This calculation is called a *hashing algorithm*, and the address is called a *hash*. Hashing is beyond the scope of this book, but it's the reason so many of the mapping operations are O(1) constant time. Sets also use hashing, because sets are essentially dictionaries with keys only instead of key-value pairs.

But keep in mind that converting a list to a set is an O(n) operation, so you don't achieve any benefit by converting a list to a set and then accessing the items in the set.

Analyzing Big O at a Glance

Once you've become familiar with performing big O analysis, you usually won't need to run through each of the steps. After a while you'll be able to just look for some telltale features in the code to quickly determine the big O order.

Keeping in mind that *n* is the size of the data the code operates on, here are some general rules you can use:

- If the code doesn't access any of the data, it's O(1).
- If the code loops over the data, it's O(n).
- If the code has two nested loops that each iterate over the data, it's O(n^2).
- Function calls don't count as one step but rather the total steps of the code inside the function. See "Big O Order of Common Function Calls" on page 242.
- If the code has a *divide and conquer* step that repeatedly halves the data, it's O(log n).
- If the code has a *divide and conquer* step that is done once per item in the data, it's an O(n log n).
- If the code goes through every possible combination of values in the *n* data, it's O(2^n), or some other exponential order.
- If the code goes through every possible permutation (that is, ordering) of the values in the data, it's O(n!).
- If the code involves sorting the data, it will be at least O(n log n).

These rules are good starting points. But they're no substitute for actual big O analysis. Keep in mind that big O order isn't the final judgment on whether code is slow, fast, or efficient. Consider the following `waitAnHour()` function:

```
import time
def waitAnHour():
    time.sleep(3600)
```

Technically, the `waitAnHour()` function is O(1) constant time. We think of constant time code as fast, yet its runtime is one hour! Does that make this code inefficient? No: it's hard to see how you could have programmed a `waitAnHour()` function that runs faster than, well, one hour.

Big O isn't a replacement for profiling your code. The point of big O notation is to give you insights as to how the code will perform under increasing amounts of input data.

Big O Doesn't Matter When n Is Small, and n Is Usually Small

Armed with the knowledge of big O notation, you might be eager to analyze every piece of code you write. Before you start using this tool to hammer every nail in sight, keep in mind that big O notation is most useful when there is a large amount of data to process. In real-world cases, the amount of data is usually small.

In those situations, coming up with elaborate, sophisticated algorithms with lower big O orders might not be worth the effort. Go programming language designer Rob Pike has five rules about programming, one of which is: "Fancy algorithms are slow when 'n' is small, and 'n' is usually small." Most software developers won't be dealing with massive data centers or complex calculations but rather more mundane programs. In these circumstances, running your code under a profiler will yield more concrete information about the code's performance than big O analysis.

Summary

The Python standard library comes with two modules for profiling: `timeit` and `cProfile`. The `timeit.timeit()` function is useful for running small snippets of code to compare the speed difference between them. The `cProfile.run()` function compiles a detailed report on larger functions and can point out any bottlenecks.

It's important to measure the performance of your code rather than make assumptions about it. Clever tricks to speed up your program might actually slow it down. Or you might spend lots of time optimizing what turns out to be an insignificant aspect of your program. Amdahl's Law captures this mathematically. The formula describes how a speed-up to one component affects the speed-up of the overall program.

Big O is the most widely used practical concept in computer science for programmers. It requires a bit of math to understand, but the underlying concept of figuring out how code slows as data grows can describe algorithms without requiring significant number crunching.

There are seven common orders of big O notation: $O(1)$, or constant time, describes code that doesn't change as the size of the data n grows; $O(\log n)$, or logarithmic time, describes code that increases by one step as the n data doubles in size; $O(n)$, or linear time, describes code that slows in proportion to the n data's growth; $O(n \log n)$, or n-log-n time, describes code that is a bit slower than $O(n)$, and many sorting algorithms have this order.

The higher orders are slower, because their runtime grows much faster than the size of their input data: $O(n^2)$, or polynomial time, describes code whose runtime increases by the square of the n input; $O(2^n)$, or exponential time, and $O(n!)$, or factorial time, orders are uncommon, but come up when combinations or permutations are involved, respectively.

Keep in mind that although big O is a helpful analysis tool, it isn't a substitute for running your code under a profiler to find out where any bottlenecks are. But an awareness of big O notation and how code slows as data grows can help you avoid writing code that is orders slower than it needs to be.

14

PRACTICE PROJECTS

So far, this book has taught you techniques for writing readable, Pythonic code. Let's put these techniques into practice by looking at the source code for two command line games: the Tower of Hanoi and Four-in-a-Row.

These projects are short and text-based to keep their scope small, but they demonstrate the principles this book outlines so far. I formatted the code using the Black tool described in "Black: The Uncompromising Code Formatter" on page 53. I chose the variable names according to the guidelines in Chapter 4. I wrote the code in a Pythonic style, as described in Chapter 6. In addition, I wrote comments and docstrings as described in Chapter 11. Because the programs are small and we haven't yet covered object-oriented programming (OOP), I wrote these two projects without the classes you'll learn more about in Chapters 15 to 17.

This chapter presents the full source code for these two projects along with a detailed breakdown of the code. These explanations aren't so much for *how* the code works (a basic understanding of Python syntax is all that's needed for that), but *why* the code was written the way it was. Still, different software developers have different opinions on how to write code and what they deem as *Pythonic*. You're certainly welcome to question and critique the source code in these projects.

After reading through a project in this book, I recommend typing the code yourself and running the programs a few times to understand how they work. Then try to reimplement the programs from scratch. Your code doesn't have to match the code in this chapter, but rewriting the code will give you a sense of the decision making and design trade-offs that programming requires.

The Tower of Hanoi

The Tower of Hanoi puzzle uses a stack of disks of different sizes. The disks have holes in their centers, so you can place them over one of three poles (Figure 14-1). To solve the puzzle, the player must move the stack of disks to one of the other poles. There are three restrictions:

1. The player can move only one disk at a time.
2. The player can only move disks to and from the top of a tower.
3. The player can never place a larger disk on top of a smaller disk.

Figure 14-1: A physical Tower of Hanoi puzzle set

Solving this puzzle is a common computer science problem used for teaching recursive algorithms. Our program won't solve this puzzle; rather, it will present the puzzle to a human player to solve. You'll find more information about the Tower of Hanoi at *https://en.wikipedia.org/wiki/Tower_of_Hanoi*.

The Output

The Tower of Hanoi program displays the towers as ASCII art by using text characters to represent the disks. It might look primitive compared to modern apps, but this approach keeps the implementation simple, because we only need print() and input() calls to interact with the user. When you run the program, the output will look something like the following. The text the player enters is in bold.

```
THE TOWER OF HANOI, by Al Sweigart al@inventwithpython.com

Move the tower of disks, one disk at a time, to another tower. Larger
disks cannot rest on top of a smaller disk.

More info at https://en.wikipedia.org/wiki/Tower_of_Hanoi

        ||           ||           ||
      @_1@           ||           ||
     @@_2@@          ||           ||
    @@@_3@@@         ||           ||
   @@@@_4@@@@        ||           ||
  @@@@@_5@@@@@       ||           ||
        A            B            C

Enter the letters of "from" and "to" towers, or QUIT.
(e.g., AB to moves a disk from tower A to tower B.)

> AC
        ||           ||           ||
        ||           ||           ||
     @@_2@@          ||           ||
    @@@_3@@@         ||           ||
   @@@@_4@@@@        ||           ||
  @@@@@_5@@@@@       ||         @_1@
        A            B            C

Enter the letters of "from" and "to" towers, or QUIT.
(e.g., AB to moves a disk from tower A to tower B.)

--snip--

        ||           ||           ||
        ||           ||         @_1@
        ||           ||        @@_2@@
        ||           ||       @@@_3@@@
        ||           ||      @@@@_4@@@@
        ||           ||     @@@@@_5@@@@@
        A            B            C

You have solved the puzzle! Well done!
```

For n disks, it takes a minimum of $2^n - 1$ moves to solve the Tower of Hanoi. So this five-disk tower requires 31 steps: AC, AB, CB, AC, BA, BC, AC, AB, CB, CA, BA, CB, AC, AB, CB, AC, BA, BC, AC, BA, CB, CA, BA,

BC, AC, AB, CB, AC, BA, BC, and finally AC. If you want a greater challenge to solve on your own, you can increase the TOTAL_DISKS variable in the program from 5 to 6.

The Source Code

Open a new file in your editor or IDE, and enter the following code. Save it as *towerofhanoi.py*.

```
"""THE TOWER OF HANOI, by Al Sweigart al@inventwithpython.com
A stack-moving puzzle game."""

import copy
import sys

TOTAL_DISKS = 5  # More disks means a more difficult puzzle.

# Start with all disks on tower A:
SOLVED_TOWER = list(range(TOTAL_DISKS, 0, -1))


def main():
    """Runs a single game of The Tower of Hanoi."""
    print(
        """THE TOWER OF HANOI, by Al Sweigart al@inventwithpython.com

Move the tower of disks, one disk at a time, to another tower. Larger
disks cannot rest on top of a smaller disk.

More info at https://en.wikipedia.org/wiki/Tower_of_Hanoi
"""
    )

    """The towers dictionary has keys "A", "B", and "C" and values
    that are lists representing a tower of disks. The list contains
    integers representing disks of different sizes, and the start of
    the list is the bottom of the tower. For a game with 5 disks,
    the list [5, 4, 3, 2, 1] represents a completed tower. The blank
    list [] represents a tower of no disks. The list [1, 3] has a
    larger disk on top of a smaller disk and is an invalid
    configuration. The list [3, 1] is allowed since smaller disks
    can go on top of larger ones."""
    towers = {"A": copy.copy(SOLVED_TOWER), "B": [], "C": []}

    while True:  # Run a single turn on each iteration of this loop.
        # Display the towers and disks:
        displayTowers(towers)

        # Ask the user for a move:
        fromTower, toTower = getPlayerMove(towers)

        # Move the top disk from fromTower to toTower:
        disk = towers[fromTower].pop()
        towers[toTower].append(disk)
```

```python
        # Check if the user has solved the puzzle:
        if SOLVED_TOWER in (towers["B"], towers["C"]):
            displayTowers(towers)  # Display the towers one last time.
            print("You have solved the puzzle! Well done!")
            sys.exit()


def getPlayerMove(towers):
    """Asks the player for a move. Returns (fromTower, toTower)."""

    while True:  # Keep asking player until they enter a valid move.
        print('Enter the letters of "from" and "to" towers, or QUIT.')
        print("(e.g., AB to moves a disk from tower A to tower B.)")
        print()
        response = input("> ").upper().strip()

        if response == "QUIT":
            print("Thanks for playing!")
            sys.exit()

        # Make sure the user entered valid tower letters:
        if response not in ("AB", "AC", "BA", "BC", "CA", "CB"):
            print("Enter one of AB, AC, BA, BC, CA, or CB.")
            continue  # Ask player again for their move.

        # Use more descriptive variable names:
        fromTower, toTower = response[0], response[1]

        if len(towers[fromTower]) == 0:
            # The "from" tower cannot be an empty tower:
            print("You selected a tower with no disks.")
            continue  # Ask player again for their move.
        elif len(towers[toTower]) == 0:
            # Any disk can be moved onto an empty "to" tower:
            return fromTower, toTower
        elif towers[toTower][-1] < towers[fromTower][-1]:
            print("Can't put larger disks on top of smaller ones.")
            continue  # Ask player again for their move.
        else:
            # This is a valid move, so return the selected towers:
            return fromTower, toTower


def displayTowers(towers):
    """Display the three towers with their disks."""

    # Display the three towers:
    for level in range(TOTAL_DISKS, -1, -1):
        for tower in (towers["A"], towers["B"], towers["C"]):
            if level >= len(tower):
                displayDisk(0)  # Display the bare pole with no disk.
            else:
                displayDisk(tower[level])  # Display the disk.
        print()
```

```
    # Display the tower labels A, B, and C:
    emptySpace = " " * (TOTAL_DISKS)
    print("{0} A{0}{0} B{0}{0} C\n".format(emptySpace))


def displayDisk(width):
    """Display a disk of the given width. A width of 0 means no disk."""
    emptySpace = " " * (TOTAL_DISKS - width)

    if width == 0:
        # Display a pole segment without a disk:
        print(f"{emptySpace}||{emptySpace}", end="")
    else:
        # Display the disk:
        disk = "@" * width
        numLabel = str(width).rjust(2, "_")
        print(f"{emptySpace}{disk}{numLabel}{disk}{emptySpace}", end="")


# If this program was run (instead of imported), run the game:
if __name__ == "__main__":
    main()
```

Run this program and play a few games to get an idea of what this program does before reading the explanation of the source code. To check for typos, copy and paste it to the online diff tool at *https://inventwithpython.com/beyond/diff/*.

Writing the Code

Let's take a closer look at the source code to see how it follows the best practices and patterns described in this book.

We'll begin at the top of the program:

```
"""THE TOWER OF HANOI, by Al Sweigart al@inventwithpython.com
A stack-moving puzzle game."""
```

The program starts with a multiline comment that serves as a docstring for the towerofhanoi module. The built-in help() function will use this information to describe the module:

```
>>> import towerofhanoi
>>> help(towerofhanoi)
Help on module towerofhanoi:

NAME
    towerofhanoi

DESCRIPTION
    THE TOWER OF HANOI, by Al Sweigart al@inventwithpython.com
    A stack-moving puzzle game.
```

```
FUNCTIONS
    displayDisk(width)
        Display a single disk of the given width.
--snip--
```

You can add more words, even paragraphs of information, to the module's docstring if you need to. I've written only a small amount here because the program is so simple.

After the module docstring are the `import` statements:

```
import copy
import sys
```

Black formats these as separate statements rather than a single one, such as `import copy, sys`. This makes the addition or removal of imported modules easier to see in version control systems, such as Git, that track changes programmers make.

Next, we define the constants this program will need:

```
TOTAL_DISKS = 5  # More disks means a more difficult puzzle.

# Start with all disks on tower A:
SOLVED_TOWER = list(range(TOTAL_DISKS, 0, -1))
```

We define these near the top of the file to group them together and make them global variables. We've written their names in capitalized snake_case to mark them as constants.

The `TOTAL_DISKS` constant indicates how many disks the puzzle has. The `SOLVED_TOWER` variable is an example of a list that contains a solved tower: it contains every disk with the largest at the bottom and the smallest at the top. We generate this value from the `TOTAL_DISKS` value, and for five disks it's `[5, 4, 3, 2, 1]`.

Notice that there are no type hints in this file. The reason is that we can infer the types of all variables, parameters, and return values from the code. For example, we've assigned the `TOTAL_DISKS` constant the integer value 5. From this, type checkers, such as Mypy, would infer that `TOTAL_DISKS` should contain integers only.

We define a `main()` function, which the program calls near the bottom of the file:

```
def main():
    """Runs a single game of The Tower of Hanoi."""
    print(
        """THE TOWER OF HANOI, by Al Sweigart al@inventwithpython.com

Move the tower of disks, one disk at a time, to another tower. Larger
disks cannot rest on top of a smaller disk.

More info at https://en.wikipedia.org/wiki/Tower_of_Hanoi
"""
    )
```

Functions can have docstrings, too. Notice the docstring for `main()` below the `def` statement. You can view this docstring by running `import towerofhanoi` and `help(towerofhanoi.main)` from the interactive shell.

Next, we write a comment that extensively describes the data structure we use to represent the tower, because it forms the core of how this program works:

```
"""The towers dictionary has keys "A", "B", and "C" and values
that are lists representing a tower of disks. The list contains
integers representing disks of different sizes, and the start of
the list is the bottom of the tower. For a game with 5 disks,
the list [5, 4, 3, 2, 1] represents a completed tower. The blank
list [] represents a tower of no disks. The list [1, 3] has a
larger disk on top of a smaller disk and is an invalid
configuration. The list [3, 1] is allowed since smaller disks
can go on top of larger ones."""
towers = {"A": copy.copy(SOLVED_TOWER), "B": [], "C": []}
```

We use the `SOLVED_TOWER` list as a *stack*, one of the simplest data structures in software development. A stack is an ordered list of values altered only through adding (also called *pushing*) or removing (also called *popping*) values from the *top* of the stack. This data structure perfectly represents the tower in our program. We can turn a Python list into a stack if we use the `append()` method for pushing and the `pop()` method for popping, and avoid altering the list in any other way. We'll treat the end of the list as the top of the stack.

Each integer in the `towers` list represents a single disk of a certain size. For example, in a game with five disks, the list [5, 4, 3, 2, 1] would represent a full stack of disks from the largest (5) at the bottom to the smallest (1) at the top.

Notice that our comment also provides examples of a valid and invalid tower stack.

Inside the `main()` function, we write an infinite loop that runs a single turn of our puzzle game:

```
while True:  # Run a single turn on each iteration of this loop.
    # Display the towers and disks:
    displayTowers(towers)

    # Ask the user for a move:
    fromTower, toTower = getPlayerMove(towers)

    # Move the top disk from fromTower to toTower:
    disk = towers[fromTower].pop()
    towers[toTower].append(disk)
```

In a single turn, the player views the current state of the towers and enters a move. The program then updates the towers data structure. We've hid the details of these tasks in the `displayTowers()` and `getPlayerMove()` functions. These descriptive function names allow the `main()` function to provide a general overview of what the program does.

The next lines check whether the player has solved the puzzle by comparing the complete tower in SOLVED_TOWER to towers["B"] and towers["C"]:

```
# Check if the user has solved the puzzle:
if SOLVED_TOWER in (towers["B"], towers["C"]):
    displayTowers(towers)  # Display the towers one last time.
    print("You have solved the puzzle! Well done!")
    sys.exit()
```

We don't compare it to towers["A"], because that pole begins with an already complete tower; a player needs to form the tower on the B or C poles to solve the puzzle. Note that we reuse SOLVED_TOWER to make the starting towers and check whether the player solved the puzzle. Because SOLVED_TOWER is a constant, we can trust that it will always have the value we assigned to it at the beginning of the source code.

The condition we use is equivalent to but shorter than SOLVED_TOWER == towers["B"] or SOLVED_TOWER == towers["C"], a Python idiom we covered in Chapter 6. If this condition is True, the player has solved the puzzle, and we end the program. Otherwise, we loop back for another turn.

The getPlayerMove() function asks the player for a disk move and validates the move against the game rules:

```
def getPlayerMove(towers):
    """Asks the player for a move. Returns (fromTower, toTower)."""
    while True:  # Keep asking player until they enter a valid move.
        print('Enter the letters of "from" and "to" towers, or QUIT.')
        print("(e.g., AB to moves a disk from tower A to tower B.)")
        print()
        response = input("> ").upper().strip()
```

We start an infinite loop that continues looping until either a return statement causes the execution to leave the loop and function or a sys.exit() call terminates the program. The first part of the loop asks the player to enter a move by specifying *from* and *to* towers.

Notice the input("> ").upper().strip() instruction that receives keyboard input from the player. The input("> ") call accepts text input from the player by presenting a > prompt. This symbol indicates that the player should enter something. If the program didn't present a prompt, the player might momentarily think the program had frozen.

We call the upper() method on the string returned from input() so it returns an uppercase form of the string. This allows the player to enter either uppercase or lowercase tower labels, such as 'a' or 'A' for tower A. In turn, the uppercase string's strip() method is called, returning a string without any whitespace on either side in case the user accidentally added a space when entering their move. This user friendliness makes our program slightly easier for players to use.

Still in the getPlayerMove() function, we check the input the user enters:

```
if response == "QUIT":
    print("Thanks for playing!")
```

```
        sys.exit()

    # Make sure the user entered valid tower letters:
    if response not in ("AB", "AC", "BA", "BC", "CA", "CB"):
        print("Enter one of AB, AC, BA, BC, CA, or CB.")
        continue  # Ask player again for their move.
```

If the user enters 'QUIT' (in any case, or even with spaces at the beginning or end of the string, due to the calls to upper() and strip()), the program terminates. We could have made getPlayerMove() return 'QUIT' to indicate to the caller that it should call sys.exit(), rather than have getPlayerMove() call sys.exit(). But this would complicate the return value of getPlayerMove(): it would return either a tuple of two strings (for the player's move) or a single 'QUIT' string. A function that returns values of a single data type is easier to understand than a function that can return values of many possible types. I discussed this in "Return Values Should Always Have the Same Data Type" on page 177.

Between the three towers, only six to-from tower combinations are possible. Despite the fact that we hardcoded all six values in the condition that checks the move, the code is much easier to read than something like len(response) != 2 or response[0] not in 'ABC' or response[1] not in 'ABC' or response[0] == response[1]. Given these circumstances, the hardcoding approach is the most straightforward.

Generally, it's considered bad practice to hardcode values such as "AB", "AC", and other values as magic values, which are valid only as long as the program has three poles. But although we might want to adjust the number of disks by changing the TOTAL_DISKS constant, it's highly unlikely that we'll add more poles to the game. Writing out every possible pole move on this line is fine.

We create two new variables, fromTower and toTower, as descriptive names for the data. They don't serve a functional purpose, but they make the code easier to read than response[0] and response[1]:

```
    # Use more descriptive variable names:
    fromTower, toTower = response[0], response[1]
```

Next, we check whether or not the selected towers constitute a legal move:

```
    if len(towers[fromTower]) == 0:
        # The "from" tower cannot be an empty tower:
        print("You selected a tower with no disks.")
        continue  # Ask player again for their move.
    elif len(towers[toTower]) == 0:
        # Any disk can be moved onto an empty "to" tower:
        return fromTower, toTower
    elif towers[toTower][-1] < towers[fromTower][-1]:
        print("Can't put larger disks on top of smaller ones.")
        continue  # Ask player again for their move.
```

If not, a `continue` statement causes the execution to move back to the beginning of the loop, which asks the player to enter their move again. Note that we check whether `toTower` is empty; if it is, we return `fromTower, toTower` to emphasize that the move was valid, because you can always put a disk on an empty pole. These first two conditions ensure that by the time the third condition is checked, `towers[toTower]` and `towers[fromTower]` won't be empty or cause an `IndexError`. We've ordered these conditions in such a way to prevent `IndexError` or additional checking.

It's important that your programs handle any invalid input from the user or potential error cases. Users might not know what to enter, or they might make typos. Similarly, files could unexpectedly go missing, or databases could crash. Your programs need to be resilient to the exceptional cases; otherwise, they'll crash unexpectedly or cause subtle bugs later on.

If none of the previous conditions are `True`, `getPlayerMove()` returns `fromTower, toTower`:

```
        else:
            # This is a valid move, so return the selected towers:
            return fromTower, toTower
```

In Python, return statements always return a single value. Although this return statement looks like it returns two values, Python actually returns a single tuple of two values, which is equivalent to `return (fromTower, toTower)`. Python programmers often omit the parentheses in this context. The parentheses don't define a tuple as much as the commas do.

Notice that the program calls the `getPlayerMove()` function only once from the `main()` function. The function doesn't save us from duplicate code, which is the most common purpose for using one. There's no reason we couldn't put all the code in `getPlayerMove()` in the `main()` function. But we can also use functions as a way to organize code into separate units, which is how we're using `getPlayerMove()`. Doing so prevents the `main()` function from becoming too long and unwieldy.

The `displayTowers()` function displays the disks on towers A, B, and C in the `towers` argument:

```
def displayTowers(towers):
    """Display the three towers with their disks."""

    # Display the three towers:
    for level in range(TOTAL_DISKS, -1, -1):
        for tower in (towers["A"], towers["B"], towers["C"]):
            if level >= len(tower):
                displayDisk(0)  # Display the bare pole with no disk.
            else:
                displayDisk(tower[level])  # Display the disk.
        print()
```

It relies on the `displayDisk()` function, which we'll cover next, to display each disk in the tower. The `for level` loop checks every possible disk for a tower, and the `for tower` loop checks towers A, B, and C.

The displayTowers() function calls displayDisk() to display each disk at a specific width, or if 0 is passed, the pole with no disk:

```
# Display the tower labels A, B, and C:
emptySpace = ' ' * (TOTAL_DISKS)
print('{0} A{0}{0} B{0}{0} C\n'.format(emptySpace))
```

We display the A, B, and C labels onscreen. The player needs this information to distinguish between the towers and to reinforce that the towers are labeled A, B, and C rather than 1, 2, and 3 or Left, Middle, and Right. I chose not to use 1, 2, and 3 for the tower labels to prevent players from confusing these numbers with the numbers used for the disks' sizes.

We set the emptySpace variable to the number of spaces to place in between each label, which in turn is based on TOTAL_DISKS, because the more disks in the game, the wider apart the poles are. Rather than use an f-string, as in print(f'{emptySpace} A{emptySpace}{emptySpace} B{emptySpace}{emptySpace} C\n'), we use the format() string method. This allows us to use the same emptySpace argument wherever {0} appears in the associated string, producing shorter and more readable code than the f-string version.

The displayDisk() function displays a single disk along with its width. If no disk is present, it displays just the pole:

```
def displayDisk(width):
    """Display a disk of the given width. A width of 0 means no disk."""
    emptySpace = ' ' * (TOTAL_DISKS - width)
    if width == 0:
        # Display a pole segment without a disk:
        print(f'{emptySpace}||{emptySpace}', end='')
    else:
        # Display the disk:
        disk = '@' * width
        numLabel = str(width).rjust(2, '_')
        print(f"{emptySpace}{disk}{numLabel}{disk}{emptySpace}", end='')
```

We represent a disk using a leading empty space, a number of @ characters equal to the disk width, two characters for the width (including an underscore if the width is a single digit), another series of @ characters, and then the trailing empty space. To display just the empty pole, all we need are the leading empty space, two pipe characters, and trailing empty space. As a result, we'll need six calls to displayDisk() with six different arguments for width to display the following tower:

```
     ||
    @_1@
   @@_2@@
  @@@_3@@@
 @@@@_4@@@@
@@@@@_5@@@@@
```

Notice how the displayTowers() and displayDisk() functions split the responsibility of displaying the towers. Although displayTowers() decides how to interpret the data structures that represent each tower, it relies on displayDisk() to actually display each disk of the tower. Breaking your program into smaller functions like this makes each part easier to test. If the program displays the disks incorrectly, the problem is likely in displayDisk(). If the disks appear in the wrong order, the problem is likely in displayTowers(). Either way, the section of code you'll have to debug will be much smaller.

To call the main() function, we use a common Python idiom:

```
# If this program was run (instead of imported), run the game:
if __name__ == '__main__':
    main()
```

Python automatically sets the __name__ variable to '__main__' if a player runs the *towerofhanoi.py* program directly. But if someone imports the program as a module using import towerofhanoi, then __name__ would be set to 'towerofhanoi'. The if __name__ == '__main__': line will call the main() function if someone runs our program, starting a game of Tower of Hanoi. But if we simply want to import the program as a module so we could, say, call the individual functions in it for unit testing, this condition will be False and main() won't be called.

Four-in-a-Row

Four-in-a-Row is a two-player, tile-dropping game. Each player tries to create a row of four of their tiles, whether horizontally, vertically, or diagonally. It's similar to the board games *Connect Four* and *Four Up*. The game uses a 7 by 6 stand-up board, and tiles drop to the lowest unoccupied space in a column. In our Four-in-a-Row game, two human players, X and O, will play against each other, as opposed to one human player against the computer.

The Output

When you run the Four-in-a-Row program in this chapter, the output will look like this:

```
Four-in-a-Row, by Al Sweigart al@inventwithpython.com

Two players take turns dropping tiles into one of seven columns, trying
to make four in a row horizontally, vertically, or diagonally.
```

```
    1234567
   +-------+
   |.......|
   |.......|
   |.......|
   |.......|
   |.......|
   |.......|
   +-------+
```

```
Player X, enter 1 to 7 or QUIT:
> 1

    1234567
  +-------+
  |.......|
  |.......|
  |.......|
  |.......|
  |.......|
  |X......|
  +-------+
Player O, enter 1 to 7 or QUIT:
--snip--
Player O, enter 1 to 7 or QUIT:
> 4

    1234567
  +-------+
  |.......|
  |.......|
  |...O...|
  |X.OO...|
  |X.XO...|
  |XOXO..X|
  +-------+
Player O has won!
```

Try to figure out the many subtle strategies you can use to get four tiles in a row while blocking your opponent from doing the same.

The Source Code

Open a new file in your editor or IDE, enter the following code, and save it as *fourinarow.py*:

```python
"""Four-in-a-Row, by Al Sweigart al@inventwithpython.com
A tile-dropping game to get four-in-a-row, similar to Connect Four."""

import sys

# Constants used for displaying the board:
EMPTY_SPACE = "."   # A period is easier to count than a space.
PLAYER_X = "X"
PLAYER_O = "O"

# Note: Update BOARD_TEMPLATE & COLUMN_LABELS if BOARD_WIDTH is changed.
BOARD_WIDTH = 7
BOARD_HEIGHT = 6
COLUMN_LABELS = ("1", "2", "3", "4", "5", "6", "7")
assert len(COLUMN_LABELS) == BOARD_WIDTH

# The template string for displaying the board:
BOARD_TEMPLATE = """
```

```
   1234567
 +-------+
 |{}{}{}{}{}{}{}|
 |{}{}{}{}{}{}{}|
 |{}{}{}{}{}{}{}|
 |{}{}{}{}{}{}{}|
 |{}{}{}{}{}{}{}|
 |{}{}{}{}{}{}{}|
 +-------+"""


def main():
    """Runs a single game of Four-in-a-Row."""
    print(
        """Four-in-a-Row, by Al Sweigart al@inventwithpython.com

Two players take turns dropping tiles into one of seven columns, trying
to make Four-in-a-Row horizontally, vertically, or diagonally.
"""
    )

    # Set up a new game:
    gameBoard = getNewBoard()
    playerTurn = PLAYER_X

    while True:  # Run a player's turn.
        # Display the board and get player's move:
        displayBoard(gameBoard)
        playerMove = getPlayerMove(playerTurn, gameBoard)
        gameBoard[playerMove] = playerTurn

        # Check for a win or tie:
        if isWinner(playerTurn, gameBoard):
            displayBoard(gameBoard)  # Display the board one last time.
            print("Player {} has won!".format(playerTurn))
            sys.exit()
        elif isFull(gameBoard):
            displayBoard(gameBoard)  # Display the board one last time.
            print("There is a tie!")
            sys.exit()

        # Switch turns to other player:
        if playerTurn == PLAYER_X:
            playerTurn = PLAYER_O
        elif playerTurn == PLAYER_O:
            playerTurn = PLAYER_X


def getNewBoard():
    """Returns a dictionary that represents a Four-in-a-Row board.

    The keys are (columnIndex, rowIndex) tuples of two integers, and the
    values are one of the "X", "O" or "." (empty space) strings."""
    board = {}
```

```python
    for rowIndex in range(BOARD_HEIGHT):
        for columnIndex in range(BOARD_WIDTH):
            board[(columnIndex, rowIndex)] = EMPTY_SPACE
    return board


def displayBoard(board):
    """Display the board and its tiles on the screen."""

    # Prepare a list to pass to the format() string method for the board
    # template. The list holds all of the board's tiles (and empty
    # spaces) going left to right, top to bottom:
    tileChars = []
    for rowIndex in range(BOARD_HEIGHT):
        for columnIndex in range(BOARD_WIDTH):
            tileChars.append(board[(columnIndex, rowIndex)])

    # Display the board:
    print(BOARD_TEMPLATE.format(*tileChars))


def getPlayerMove(playerTile, board):
    """Let a player select a column on the board to drop a tile into.

    Returns a tuple of the (column, row) that the tile falls into."""
    while True:  # Keep asking player until they enter a valid move.
        print(f"Player {playerTile}, enter 1 to {BOARD_WIDTH} or QUIT:")
        response = input("> ").upper().strip()

        if response == "QUIT":
            print("Thanks for playing!")
            sys.exit()

        if response not in COLUMN_LABELS:
            print(f"Enter a number from 1 to {BOARD_WIDTH}.")
            continue  # Ask player again for their move.

        columnIndex = int(response) - 1  # -1 for 0-based column indexes.

        # If the column is full, ask for a move again:
        if board[(columnIndex, 0)] != EMPTY_SPACE:
            print("That column is full, select another one.")
            continue  # Ask player again for their move.

        # Starting from the bottom, find the first empty space.
        for rowIndex in range(BOARD_HEIGHT - 1, -1, -1):
            if board[(columnIndex, rowIndex)] == EMPTY_SPACE:
                return (columnIndex, rowIndex)


def isFull(board):
    """Returns True if the `board` has no empty spaces, otherwise
    returns False."""
    for rowIndex in range(BOARD_HEIGHT):
        for columnIndex in range(BOARD_WIDTH):
```

```
                    if board[(columnIndex, rowIndex)] == EMPTY_SPACE:
                        return False  # Found an empty space, so return False.
            return True  # All spaces are full.


def isWinner(playerTile, board):
    """Returns True if `playerTile` has four tiles in a row on `board`,
    otherwise returns False."""

    # Go through the entire board, checking for four-in-a-row:
    for columnIndex in range(BOARD_WIDTH - 3):
        for rowIndex in range(BOARD_HEIGHT):
            # Check for four-in-a-row going across to the right:
            tile1 = board[(columnIndex, rowIndex)]
            tile2 = board[(columnIndex + 1, rowIndex)]
            tile3 = board[(columnIndex + 2, rowIndex)]
            tile4 = board[(columnIndex + 3, rowIndex)]
            if tile1 == tile2 == tile3 == tile4 == playerTile:
                return True

    for columnIndex in range(BOARD_WIDTH):
        for rowIndex in range(BOARD_HEIGHT - 3):
            # Check for four-in-a-row going down:
            tile1 = board[(columnIndex, rowIndex)]
            tile2 = board[(columnIndex, rowIndex + 1)]
            tile3 = board[(columnIndex, rowIndex + 2)]
            tile4 = board[(columnIndex, rowIndex + 3)]
            if tile1 == tile2 == tile3 == tile4 == playerTile:
                return True

    for columnIndex in range(BOARD_WIDTH - 3):
        for rowIndex in range(BOARD_HEIGHT - 3):
            # Check for four-in-a-row going right-down diagonal:
            tile1 = board[(columnIndex, rowIndex)]
            tile2 = board[(columnIndex + 1, rowIndex + 1)]
            tile3 = board[(columnIndex + 2, rowIndex + 2)]
            tile4 = board[(columnIndex + 3, rowIndex + 3)]
            if tile1 == tile2 == tile3 == tile4 == playerTile:
                return True

            # Check for four-in-a-row going left-down diagonal:
            tile1 = board[(columnIndex + 3, rowIndex)]
            tile2 = board[(columnIndex + 2, rowIndex + 1)]
            tile3 = board[(columnIndex + 1, rowIndex + 2)]
            tile4 = board[(columnIndex, rowIndex + 3)]
            if tile1 == tile2 == tile3 == tile4 == playerTile:
                return True
    return False


# If this program was run (instead of imported), run the game:
if __name__ == "__main__":
    main()
```

Run this program and play a few games to get an idea of what this program does before reading the explanation of the source code. To check for typos, copy and paste it to the online diff tool at *https://inventwithpython.com/beyond/diff/*.

Writing the Code

Let's look at the program's source code, as we did for the Tower of Hanoi program. Once again, I formatted this code using Black with a line limit of 75 characters.

We'll begin at the top of the program:

```
"""Four-in-a-Row, by Al Sweigart al@inventwithpython.com
A tile-dropping game to get four-in-a-row, similar to Connect Four."""

import sys

# Constants used for displaying the board:
EMPTY_SPACE = "."  # A period is easier to count than a space.
PLAYER_X = "X"
PLAYER_O = "O"
```

We start the program with a docstring, module imports, and constant assignments, as we did in the Tower of Hanoi program. We define the PLAYER_X and PLAYER_O constants so we don't have to use the strings "X" and "O" throughout the program, making errors easier to catch. If we enter a typo while using the constants, such as PLAYER_XX, Python will raise NameError, instantly pointing out the problem. But if we make a typo with the "X" character, such as "XX" or "Z", the resulting bug might not be immediately obvious. As explained in "Magic Numbers" on page 71, using constants instead of the string value directly provides not just a description, but also an early warning for any typos in your source code.

Constants shouldn't change while the program runs. But the programmer can update their values in future versions of the program. For this reason, we make a note telling programmers that they should update the BOARD_TEMPLATE and COLUMN_LABELS constants, described later, if they change the value of BOARD_WIDTH:

```
# Note: Update BOARD_TEMPLATE & COLUMN_LABELS if BOARD_WIDTH is changed.
BOARD_WIDTH = 7
BOARD_HEIGHT = 6
```

Next, we create the COLUMN_LABELS constant:

```
COLUMN_LABELS = ("1", "2", "3", "4", "5", "6", "7")
assert len(COLUMN_LABELS) == BOARD_WIDTH
```

We'll use this constant later to ensure the player selects a valid column. Note that if we ever set BOARD_WIDTH to a value other than 7, we'll have to add labels to or remove labels from the COLUMN_LABELS tuple. I could have avoided this by generating the value of COLUMN_LABELS based on BOARD_WIDTH with code

like this: `COLUMN_LABELS = tuple([str(n) for n in range(1, BOARD_WIDTH + 1)])`.
But `COLUMN_LABELS` is unlikely to change in the future, because the standard
Four-in-a-Row game is played on a 7 by 6 board, so I decided to write out an
explicit tuple value.

Sure, this hardcoding is a code smell, as described in "Magic Numbers" on page 71, but it's more readable than the alternative. Also, the
assert statement warns us about changing `BOARD_WIDTH` without updating
`COLUMN_LABELS`.

As with Tower of Hanoi, the Four-in-a-Row program uses ASCII art to
draw the game board. The following lines are a single assignment statement
with a multiline string:

```
# The template string for displaying the board:
BOARD_TEMPLATE = """
   1234567
  +-------+
  |{}{}{}{}{}{}{}|
  |{}{}{}{}{}{}{}|
  |{}{}{}{}{}{}{}|
  |{}{}{}{}{}{}{}|
  |{}{}{}{}{}{}{}|
  |{}{}{}{}{}{}{}|
  +-------+"""
```

This string contains braces (`{}`) that the `format()` string method will
replace with the board's contents. (The `displayBoard()` function, explained
later, will handle this.) Because the board consists of seven columns and
six rows, we use seven brace pairs `{}` in each of the six rows to represent
every slot. Note that just like `COLUMN_LABELS`, we're technically hardcoding
the board to create a set number of columns and rows. If we ever change
`BOARD_WIDTH` or `BOARD_HEIGHT` to new integers, we'll have to update the multiline string in `BOARD_TEMPLATE` as well.

We could have written code to generate `BOARD_TEMPLATE` based on the
`BOARD_WIDTH` and `BOARD_HEIGHT` constants, like so:

```
BOARD_EDGE = "     +" + ("-" * BOARD_WIDTH) + "+"
BOARD_ROW = "     |" + ("{}" * BOARD_WIDTH) + "|\n"
BOARD_TEMPLATE = "\n    " + "".join(COLUMN_LABELS) + "\n" + BOARD_EDGE + "\n"
+ (BOARD_ROW * BOARD_WIDTH) + BOARD_EDGE
```

But this code is not as readable as a simple multiline string, and we're
unlikely to change the game board's size anyway, so we'll use the simple
multiline string.

We begin writing the `main()` function, which will call all the other functions we've made for this game:

```
def main():
    """Runs a single game of Four-in-a-Row."""
    print(
        """Four-in-a-Row, by Al Sweigart al@inventwithpython.com
```

Two players take turns dropping tiles into one of seven columns, trying
to make four-in-a-row horizontally, vertically, or diagonally.
"""
)

    # Set up a new game:
    gameBoard = getNewBoard()
    playerTurn = PLAYER_X
```

We give the main() function a docstring, viewable with the built-in help()
function. The main() function also prepares the game board for a new game
and chooses the first player.

Inside the main() function is an infinite loop:

```
    while True:  # Run a player's turn.
        # Display the board and get player's move:
        displayBoard(gameBoard)
        playerMove = getPlayerMove(playerTurn, gameBoard)
        gameBoard[playerMove] = playerTurn
```

Each iteration of this loop consists of a single turn. First, we display the
game board to the player. Second, the player selects a column to drop a tile
in, and third, we update the game board data structure.

Next, we evaluate the results of the player's move:

```
        # Check for a win or tie:
        if isWinner(playerTurn, gameBoard):
            displayBoard(gameBoard)  # Display the board one last time.
            print("Player {} has won!".format(playerTurn))
            sys.exit()
        elif isFull(gameBoard):
            displayBoard(gameBoard)  # Display the board one last time.
            print("There is a tie!")
            sys.exit()
```

If the player made a winning move, isWinner() returns True and the
game ends. If the player filled the board and there is no winner, isFull()
returns True and the game ends. Note that instead of calling sys.exit(),
we could have used a simple break statement. This would have caused the
execution to break out of the while loop, and because there is no code in
the main() function after this loop, the function would return to the main()
call at the bottom of the program, causing the program to end. But I opted
to use sys.exit() to make it clear to programmers reading the code that the
program will immediately terminate.

If the game hasn't ended, the following lines set playerTurn to the other
player:

```
        # Switch turns to other player:
        if playerTurn == PLAYER_X:
            playerTurn = PLAYER_O
        elif playerTurn == PLAYER_O:
            playerTurn = PLAYER_X
```
```

Notice that I could have made the `elif` statement into a simple `else` statement without a condition. But recall the Zen of Python tenet that *explicit is better than implicit.* This code *explicitly* says that if it's player O's turn now, it will be player X's turn next. The alternative would have just said that if it's not player X's turn now, it will be player X's turn next. Even though `if` and `else` statements are a natural fit with Boolean conditions, the `PLAYER_X` and `PLAYER_O` values aren't the same as `True`, and `False`: `not PLAYER_X` is not the same as `PLAYER_O`. Therefore, it's helpful to be direct when checking the value of `playerTurn`.

Alternatively, I could have performed the same actions in a one-liner:

```
playerTurn = {PLAYER_X: PLAYER_O, PLAYER_O: PLAYER_X}[ playerTurn]
```

This line uses the dictionary trick mentioned in "Use Dictionaries Instead of a `switch` Statement" on page 101. But like many one-liners, it's not as readable as a direct `if` and `elif` statement.

Next, we define the `getNewBoard()` function:

```
def getNewBoard():
    """Returns a dictionary that represents a Four-in-a-Row board.

    The keys are (columnIndex, rowIndex) tuples of two integers, and the
    values are one of the "X", "O" or "." (empty space) strings."""
    board = {}
    for rowIndex in range(BOARD_HEIGHT):
        for columnIndex in range(BOARD_WIDTH):
            board[(columnIndex, rowIndex)] = EMPTY_SPACE
    return board
```

This function returns a dictionary that represents a Four-in-a-Row board. It has (`columnIndex`, `rowIndex`) tuples for keys (where `columnIndex` and `rowIndex` are integers), and the `'X'`, `'O'`, or `'.'` character for the tile at each place on the board. We store these strings in `PLAYER_X`, `PLAYER_O`, and `EMPTY_SPACE`, respectively.

Our Four-in-a-Row game is rather simple, so using a dictionary to represent the game board is a suitable technique. Still, we could have used an OOP approach instead. We'll explore OOP in Chapters 15 through 17.

The `displayBoard()` function takes a game board data structure for the board argument and displays the board onscreen using the `BOARD_TEMPLATE` constant:

```
def displayBoard(board):
    """Display the board and its tiles on the screen."""

    # Prepare a list to pass to the format() string method for the board
    # template. The list holds all of the board's tiles (and empty
    # spaces) going left to right, top to bottom:
    tileChars = []
```

Recall that the BOARD_TEMPLATE is a multiline string with several brace pairs. When we call the format() method on BOARD_TEMPLATE, these braces will be replaced by the arguments passed to format().

The tileChars variable will contain a list of these arguments. We start by assigning it a blank list. The first value in tileChars will replace the first pair of braces in BOARD_TEMPLATE, the second value will replace the second pair, and so on. Essentially, we're creating a list of the values from the board dictionary:

```
for rowIndex in range(BOARD_HEIGHT):
    for columnIndex in range(BOARD_WIDTH):
        tileChars.append(board[(columnIndex, rowIndex)])

# Display the board:
print(BOARD_TEMPLATE.format(*tileChars))
```

These nested for loops iterate over every possible row and column on the board, appending them to the list in tileChars. Once these loops have finished, we pass the values in the tileChars list as individual arguments to the format() method using the star * prefix. "Using * to Create Variadic Functions" section on page 167 explained how to use this syntax to treat the values in a list as separate function arguments: the code print(*['cat', 'dog', 'rat']) is equivalent to print('cat', 'dog', 'rat'). We need the star because the format() method expects one argument for every brace pair, not a single list argument.

Next, we write the getPlayerMove() function:

```
def getPlayerMove(playerTile, board):
    """Let a player select a column on the board to drop a tile into.

    Returns a tuple of the (column, row) that the tile falls into."""
    while True:  # Keep asking player until they enter a valid move.
        print(f"Player {playerTile}, enter 1 to {BOARD_WIDTH} or QUIT:")
        response = input("> ").upper().strip()

        if response == "QUIT":
            print("Thanks for playing!")
            sys.exit()
```

The function begins with an infinite loop that waits for the player to enter a valid move. This code resembles the getPlayerMove() function in the Tower of Hanoi program. Note that the print() call at the start of the while loop uses an f-string so we don't have to change the message if we update BOARD_WIDTH.

We check that the player's response is a column; if it isn't, the continue statement moves the execution back to the start of the loop to ask the player for a valid move:

```
        if response not in COLUMN_LABELS:
            print(f"Enter a number from 1 to {BOARD_WIDTH}.")
            continue  # Ask player again for their move.
```

We could have written this input validation condition as not response. isdecimal() or spam < 1 or spam > BOARD_WIDTH, but it's simpler to just use response not in COLUMN_LABELS.

Next, we need to find out which row a tile dropped in the player's selected column would land on:

```
columnIndex = int(response) - 1  # -1 for 0-based column indexes.

# If the column is full, ask for a move again:
if board[(columnIndex, 0)] != EMPTY_SPACE:
    print("That column is full, select another one.")
    continue  # Ask player again for their move.
```

The board displays the column labels 1 to 7 onscreen. But the (columnIndex, rowIndex) indexes on the board use 0-based indexing, so they range from 0 to 6. To solve this discrepancy, we convert the string values '1' to '7' to the integer values 0 to 6.

The row indexes start at 0 at the top of the board and increase to 6 at the bottom of the board. We check the top row in the selected column to see whether it's occupied. If so, this column is completely full and the continue statement moves the execution back to the start of the loop to ask the player for another move.

If the column isn't full, we need to find the lowest unoccupied space for the tile to land on:

```
# Starting from the bottom, find the first empty space.
for rowIndex in range(BOARD_HEIGHT - 1, -1, -1):
    if board[(columnIndex, rowIndex)] == EMPTY_SPACE:
        return (columnIndex, rowIndex)
```

This for loop starts at the bottom row index, BOARD_HEIGHT - 1 or 6, and moves up until it finds the first empty space. The function then returns the indexes of the lowest empty space.

Anytime the board is full, the game ends in a tie:

```
def isFull(board):
    """Returns True if the `board` has no empty spaces, otherwise
    returns False."""
    for rowIndex in range(BOARD_HEIGHT):
        for columnIndex in range(BOARD_WIDTH):
            if board[(columnIndex, rowIndex)] == EMPTY_SPACE:
                return False  # Found an empty space, so return False.
    return True  # All spaces are full.
```

The isFull() function uses a pair of nested for loops to iterate over every place on the board. If it finds a single empty space, the board isn't full, and the function returns False. If the execution makes it through both loops, the isFull() function found no empty space, so it returns True.

The isWinner() function checks whether a player has won the game:

```
def isWinner(playerTile, board):
    """Returns True if `playerTile` has four tiles in a row on `board`,
    otherwise returns False."""

    # Go through the entire board, checking for four-in-a-row:
    for columnIndex in range(BOARD_WIDTH - 3):
        for rowIndex in range(BOARD_HEIGHT):
            # Check for four-in-a-row going across to the right:
            tile1 = board[(columnIndex, rowIndex)]
            tile2 = board[(columnIndex + 1, rowIndex)]
            tile3 = board[(columnIndex + 2, rowIndex)]
            tile4 = board[(columnIndex + 3, rowIndex)]
            if tile1 == tile2 == tile3 == tile4 == playerTile:
                return True
```

This function returns True if playerTile appears four times in a row horizontally, vertically, or diagonally. To figure out whether the condition is met, we have to check every set of four adjacent spaces on the board. We'll use a series of nested for loops to do this.

The (columnIndex, rowIndex) tuple represents a starting point. We check the starting point and the three spaces to the right of it for the playerTile string. If the starting space is (columnIndex, rowIndex), the space to its right will be (columnIndex + 1, rowIndex), and so on. We'll save the tiles in these four spaces to the variables tile1, tile2, tile3, and tile4. If all of these variables have the same value as playerTile, we've found a four-in-a-row, and the isWinner() function returns True.

In "Variables with Numeric Suffixes" on page 76, I mentioned that variable names with sequential numeric suffixes (like tile1 through tile4 in this game) are often a code smell that indicates you should use a single list instead. But in this context, these variable names are fine. We don't need to replace them with a list, because the Four-in-a-Row program will always require exactly four of these tile variables. Remember that a code smell doesn't necessarily indicate a problem; it only means we should take a second look and confirm that we've written our code in the most readable way. In this case, using a list would make our code more complicated, and it wouldn't add any benefit, so we'll stick to using tile1, tile2, tile3, and tile4.

We use a similar process to check for vertical four-in-a-row tiles:

```
    for columnIndex in range(BOARD_WIDTH):
        for rowIndex in range(BOARD_HEIGHT - 3):
            # Check for four-in-a-row going down:
            tile1 = board[(columnIndex, rowIndex)]
            tile2 = board[(columnIndex, rowIndex + 1)]
            tile3 = board[(columnIndex, rowIndex + 2)]
            tile4 = board[(columnIndex, rowIndex + 3)]
            if tile1 == tile2 == tile3 == tile4 == playerTile:
                return True
```

Next, we check for four-in-a-row tiles in a diagonal pattern going down and to the right; then we check for four-in-a-row tiles in a diagonal pattern going down and to the left:

```
for columnIndex in range(BOARD_WIDTH - 3):
    for rowIndex in range(BOARD_HEIGHT - 3):
        # Check for four-in-a-row going right-down diagonal:
        tile1 = board[(columnIndex, rowIndex)]
        tile2 = board[(columnIndex + 1, rowIndex + 1)]
        tile3 = board[(columnIndex + 2, rowIndex + 2)]
        tile4 = board[(columnIndex + 3, rowIndex + 3)]
        if tile1 == tile2 == tile3 == tile4 == playerTile:
            return True

        # Check for four-in-a-row going left-down diagonal:
        tile1 = board[(columnIndex + 3, rowIndex)]
        tile2 = board[(columnIndex + 2, rowIndex + 1)]
        tile3 = board[(columnIndex + 1, rowIndex + 2)]
        tile4 = board[(columnIndex, rowIndex + 3)]
        if tile1 == tile2 == tile3 == tile4 == playerTile:
            return True
```

This code is similar to the horizontal four-in-a-row checks, so I won't repeat the explanation here. If all the checks for four-in-a-row tiles fail to find any, the function returns False to indicate that playerTile is not a winner on this board:

```
return False
```

The only task left is to call the main() function:

```
# If this program was run (instead of imported), run the game:
if __name__ == '__main__':
    main()
```

Once again, we use a common Python idiom that will call main() if *fourinarow.py* is run directly but not if *fourinarow.py* is imported as a module.

Summary

The Tower of Hanoi puzzle game and Four-in-a-Row game are short programs, but by following the practices in this book, you can ensure that their code is readable and easy to debug. These programs follow several good practices: they've been automatically formatted with Black, use docstrings to describe the module and functions, and place the constants near the top of the file. They limit the variables, function parameters, and function return values to a single data type so type hinting, although a beneficial form of additional documentation, is unnecessary.

In the Tower of Hanoi, we represent the three towers as a dictionary with keys 'A', 'B', and 'C' whose values are lists of integers. This works, but if our program were any larger or more complicated, it would be a good idea to represent this data using a class. Classes and OOP techniques weren't used in this chapter because I don't cover OOP until Chapters 15 through 17. But keep in mind that it's perfectly valid to use a class for this data structure. The towers render as ASCII art onscreen, using text characters to show each disk of the towers.

The Four-in-a-Row game uses ASCII art to display a representation of the game board. We display this using a multiline string stored in the BOARD_TEMPLATE constant. This string has 42 brace pairs {} to display each space on the 7 by 6 board. We use braces so the format() string method can replace them with the tile at that space. This way, it's more obvious how the BOARD_TEMPLATE string produces the game board as it appears onscreen.

Although their data structures differ, these two programs share many similarities. They both render their data structures onscreen, ask the player for input, validate that input, and then use it to update their data structures before looping back to the beginning. But there are many different ways we could have written code to carry out these actions. What makes code readable is ultimately a subjective opinion rather than an objective measure of how closely it adheres to some list of rules. The source code in this chapter shows that although we should always give any code smells a second look, not all code smells indicate a problem that we need to fix. Code readability is more important than mindlessly following a "zero code smells" policy for your programs.

PART 3

OBJECT-ORIENTED PYTHON

15

OBJECT-ORIENTED PROGRAMMING AND CLASSES

OOP is a programming language feature that allows you to group variables and functions together into new data types, called *classes*, from which you can create objects. By organizing your code into classes, you can break down a monolithic program into smaller parts that are easier to understand and debug.

For small programs, OOP doesn't add organization so much as it adds bureaucracy. Although some languages, such as Java, require you to organize all your code into classes, Python's OOP features are optional. Programmers can take advantage of classes if they need them or ignore them if they don't. Python core developer Jack Diederich's PyCon 2012 talk, "Stop Writing Classes" (*https://youtu.be/o9pEzgHorH0/*), points out many cases where programmers write classes when a simpler function or module would have worked better.

That said, as a programmer, you should be familiar with the basics of what classes are and how they work. In this chapter, you'll learn what classes are, why they're used in programs, and the syntax and programming concepts behind them. OOP is a broad topic, and this chapter acts only as an introduction.

Real-World Analogy: Filling Out a Form

You've most likely had to fill out paper or electronic forms numerous times in your life: for doctor's visits, for online purchases, or to RSVP to a wedding. Forms exist as a uniform way for another person or organization to collect the information they need about you. Different forms ask for different kinds of information. You would report a sensitive medical condition on a doctor's form, and you would report any guests you're bringing on a wedding RSVP, but not the other way around.

In Python, *class*, *type*, and *data type* have the same meaning. Like a paper or electronic form, a *class* is a blueprint for Python *objects* (also called *instances*), which contain the data that represents a noun. This noun could be a doctor's patient, an ecommerce purchase, or a wedding guest. Classes are like a blank form template, and the objects created from that class are like filled-out forms that contain actual data about the kind of thing the form represents. For example, in Figure 15-1, the RSVP response form is like a class, whereas the filled-out RSVP is like an object.

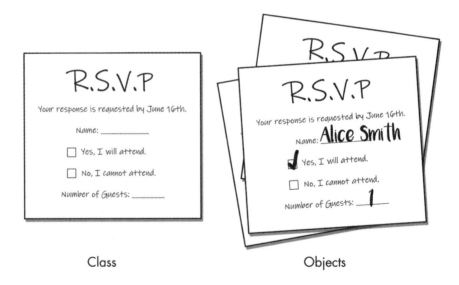

Figure 15-1: Wedding RSVP form templates are like classes, and filled-out forms are like objects.

You can also think of classes and objects as spreadsheets, as in Figure 15-2.

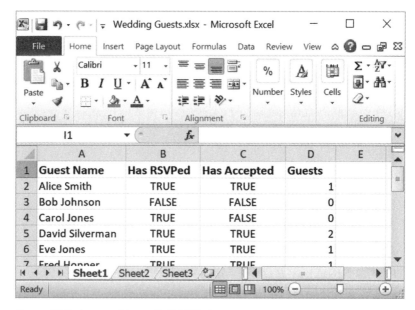

Figure 15-2: A spreadsheet of all RSVP data

The column headers would make up the class, and the individual rows would each make up an object.

Classes and objects are often talked about as data models of items in the real world, but don't confuse the map for the territory. What goes into the class depends on what the program needs to do. Figure 15-3 shows some objects of different classes that represent the same real-world person, and other than the person's name, they store completely different information.

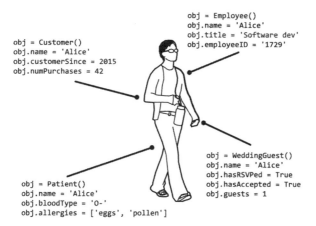

Figure 15-3: Four objects made from different classes that represent the same real-world person, depending on what the software application needs to know about the person

Also, the information contained in your classes should depend on your program's needs. Many OOP tutorials use a Car class as their basic example without noting that what goes into a class depends entirely on the kind of software you're writing. There's no such thing as a generic Car class that would *obviously* have a honkHorn() method or a numberOfCupholders attribute just because those are characteristics real-world cars have. Your program might be for a car dealership web app, a car racing video game, or a road traffic simulation. The car dealership web app's Car class might have milesPerGallon or manufacturersSuggestedRetailPrice attributes (just as a car dealership's spreadsheets might use these as columns). But the video game and road traffic simulation wouldn't have these attributes, because this information isn't relevant to them. The video game's Car class might have an explodeWithLargeFireball() method, but the car dealership and traffic simulation, hopefully, would not.

Creating Objects from Classes

You've already used classes and objects in Python, even if you haven't created classes yourself. Consider the datetime module, which contains a class named date. Objects of the datetime.date class (also simply called datetime. date objects or date objects) represent a specific date. Enter the following in the interactive shell to create an object of the datetime.date class:

```
>>> import datetime
>>> birthday = datetime.date(1999, 10, 31) # Pass the year, month, and day.
>>> birthday.year
1999
>>> birthday.month
10
>>> birthday.day
31
>>> birthday.weekday() # weekday() is a method; note the parentheses.
6
```

Attributes are variables associated with objects. The call to datetime.date() creates a new date object, initialized with the arguments 1999, 10, 31 so the object represents the date October 31, 1999. We assign these arguments as the date class's year, month, and day attributes, which all date objects have.

With this information, the class's weekday() method can calculate the day of the week. In this example, it returns 6 for Sunday, because according to Python's online documentation, the return value of weekday() is an integer that starts at 0 for Monday and goes to 6 for Sunday. The documentation lists several other methods that objects of the date class have. Even though the date object contains multiple attributes and methods, it's still a single object that you can store in a variable, such as birthday in this example.

Creating a Simple Class: WizCoin

Let's create a WizCoin class, which represents a number of coins in a fictional wizard currency. In this currency, the denominations are knuts, sickles (worth 29 knuts), and galleons (worth 17 sickles or 493 knuts). Keep in mind that the objects in the WizCoin class represent a quantity of coins, not an amount of money. For example, it will inform you that you're holding five quarters and one dime rather than $1.35.

In a new file named *wizcoin.py*, enter the following code to create the WizCoin class. Note that the _init_ method name has two underscores before and after init (we'll discuss _init_ in "Methods, __init__(), and self" later in this chapter):

```
❶ class WizCoin:
❷     def __init__(self, galleons, sickles, knuts):
           """Create a new WizCoin object with galleons, sickles, and knuts."""
           self.galleons = galleons
           self.sickles  = sickles
           self.knuts    = knuts
           # NOTE: __init__() methods NEVER have a return statement.

❸     def value(self):
           """The value (in knuts) of all the coins in this WizCoin object."""
           return (self.galleons * 17 * 29) + (self.sickles * 29) + (self.knuts)

❹     def weightInGrams(self):
           """Returns the weight of the coins in grams."""
           return (self.galleons * 31.103) + (self.sickles * 11.34) + (self.knuts
    * 5.0)
```

This program defines a new class called WizCoin using a class statement ❶. Creating a class creates a new type of object. Using a class statement to define a class is similar to def statements that define new functions. Inside the block of code following the class statement are the definitions for three methods: __init__() (short for *initializer*) ❷, value() ❸, and weightInGrams() ❹. Note that all methods have a first parameter named self, which we'll explore in the next section.

As a convention, module names (like wizcoin in our *wizcoin.py* file) are lowercase, whereas class names (like WizCoin) begin with an uppercase letter. Unfortunately, some classes in the Python Standard Library, such as date, don't follow this convention.

To practice creating new objects of the WizCoin class, enter the following source code in a separate file editor window and save the file as *wcexample1.py* in the same folder as *wizcoin.py*:

```
import wizcoin

❶ purse = wizcoin.WizCoin(2, 5, 99) # The ints are passed to __init__().
  print(purse)
  print('G:', purse.galleons, 'S:', purse.sickles, 'K:', purse.knuts)
  print('Total value:', purse.value())
  print('Weight:', purse.weightInGrams(), 'grams')
```

```
    print()

❷ coinJar = wizcoin.WizCoin(13, 0, 0) # The ints are passed to __init__().
  print(coinJar)
  print('G:', coinJar.galleons, 'S:', coinJar.sickles, 'K:', coinJar.knuts)
  print('Total value:', coinJar.value())
  print('Weight:', coinJar.weightInGrams(), 'grams')
```

The calls to WizCoin() ❶ ❷ create a WizCoin object and run the code in the __init__() method for them. We pass in three integers as arguments to WizCoin(), which are forwarded to the parameters of __init__(). These arguments are assigned to the object's self.galleons, self.sickles, and self.knuts attributes. Note that, just as the time.sleep() function requires you to first import the time module and put time. before the function name, we must also import wizcoin and put wizcoin. before the WizCoin() function name.

When you run this program, the output will look something like this:

```
<wizcoin.WizCoin object at 0x000002136F138080>
G: 2 S: 5 K: 99
Total value: 1230
Weight: 613.906 grams

<wizcoin.WizCoin object at 0x000002136F138128>
G: 13 S: 0 K: 0
Total value: 6409
Weight: 404.339 grams
```

If you get an error message, such as ModuleNotFoundError: No module named 'wizcoin', check to make sure that your file is named *wizcoin.py* and that it's in the same folder as *wcexample1.py*.

The WizCoin objects don't have useful string representations, so printing purse and coinJar displays a memory address in between angle brackets. (You'll learn how to change this in Chapter 17.)

Just as we can call the lower() string method on a string object, we can call the value() and weightInGrams() methods on the WizCoin objects we've assigned to the purse and coinJar variables. These methods calculate values based on the object's galleons, sickles, and knuts attributes.

Classes and OOP can lead to more *maintainable* code—that is, code that is easier to read, modify, and extend in the future. Let's explore this class's methods and attributes in more detail.

Methods, __init__(), and self

Methods are functions associated with objects of a particular class. Recall that lower() is a string method, meaning that it's called on string objects. You can call lower() on a string, as in 'Hello'.lower(), but you can't call it on a list, such as ['dog', 'cat'].lower(). Also, notice that methods come after the object: the correct code is 'Hello'.lower(), not lower('Hello'). Unlike a method like lower(), a function like len() is not associated with a single data type; you can pass strings, lists, dictionaries, and many other types of objects to len().

As you saw in the previous section, we create objects by calling the class name as a function. This function is referred to as a *constructor function* (or *constructor*, or abbreviated as *ctor*, pronounced "see-tore") because it constructs a new object. We also say the constructor *instantiates* a new instance of the class.

Calling the constructor causes Python to create the new object and then run the __init__() method. Classes aren't required to have an __init__() method, but they almost always do. The __init__() method is where you commonly set the initial values of attributes. For example, recall that the __init__() method of WizCoin looks like the following:

```
def __init__(self, galleons, sickles, knuts):
        """Create a new WizCoin object with galleons, sickles, and knuts."""
        self.galleons = galleons
        self.sickles  = sickles
        self.knuts    = knuts
    # NOTE: __init__() methods NEVER have a return statement.
```

When the *wcexample1.py* program calls WizCoin(2, 5, 99), Python creates a new WizCoin object and then passes three arguments (2, 5, and 99) to an __init__() call. But the __init__() method has four parameters: self, galleons, sickles, and knuts. The reason is that all methods have a first parameter named self. When a method is called on an object, the object is automatically passed in for the self parameter. The rest of the arguments are assigned to parameters normally. If you see an error message, such as TypeError: __init__() takes 3 positional arguments but 4 were given, you've probably forgotten to add the self parameter to the method's def statement.

You don't have to name a method's first parameter self; you can name it anything. But using self is conventional, and choosing a different name will make your code less readable to other Python programmers. When you're reading code, the presence of self as the first parameter is the quickest way you can distinguish methods from functions. Similarly, if your method's code never needs to use the self parameter, it's a sign that your method should probably just be a function.

The 2, 5, and 99 arguments of WizCoin(2, 5, 99) aren't automatically assigned to the new object's attributes; we need the three assignment statements in __init__() to do this. Often, the __init__() parameters are named the same as the attributes, but the presence of self in self.galleons indicates that it's an attribute of the object, whereas galleons is a parameter. This storing of the constructor's arguments in the object's attributes is a common task for a class's __init__() method. The datetime.date() call in the previous section did a similar task except the three arguments we passed were for the newly created date object's year, month, and day attributes.

You've previously called the int(), str(), float(), and bool() functions to convert between data types, such as str(3.1415) returning the string value '3.1415' based on the float value 3.1415. Previously, we described these as functions, but int, str, float, and bool are actually classes, and the int(), str(), float(), and bool() functions are constructor functions that return new integer, string, float, and Boolean objects. Python's style guide recommends using capitalized camelcase for your class names (like WizCoin), although many of Python's built-in classes don't follow this convention.

Note that calling the WizCoin() construction function returns the new WizCoin object, but the __init__() method never has a return statement with a return value. Adding a return value causes this error: TypeError: __init__() should return None.

Attributes

Attributes are variables associated with an object. The Python documentation describes attributes as "any name following a dot." For example, consider the birthday.year expression in the previous section. The year attribute is a name following a dot.

Every object has its own set of attributes. When the *wcexample1.py* program created two WizCoin objects and stored them in the purse and coinJar variables, their attributes had different values. You can access and set these attributes just like any variable. To practice setting attributes, open a new file editor window and enter the following code, saving it as *wcexample2.py* in the same folder as the *wizcoin.py* file:

```
import wizcoin

change = wizcoin.WizCoin(9, 7, 20)
print(change.sickles) # Prints 7.
change.sickles += 10
print(change.sickles) # Prints 17.

pile = wizcoin.WizCoin(2, 3, 31)
print(pile.sickles) # Prints 3.
pile.someNewAttribute = 'a new attr' # A new attribute is created.
print(pile.someNewAttribute)
```

When you run this program, the output looks like this:

```
7
17
3
a new attr
```

You can think of an object's attributes as similar to a dictionary's keys. You can read and modify their associated values and assign an object new attributes. Technically, methods are considered attributes of a class, as well.

Private Attributes and Private Methods

In languages such as C++ or Java, attributes can be marked as having *private access*, which means the compiler or interpreter only lets code inside the class's methods access or modify the attributes of objects of that class. But in Python, this enforcement doesn't exist. All attributes and methods are effectively *public access*: code outside of the class can access and modify any attribute in any object of that class.

But private access is useful. For example, objects of a BankAccount class could have a balance attribute that only methods of the BankAccount class should have access to. For those reasons, Python's convention is to start *private* attribute or method names with a single underscore. Technically, there is nothing to stop code outside the class from accessing private attributes and methods, but it's a best practice to let only the class's methods access them.

Open a new file editor window, enter the following code, and save it as *privateExample.py*. In it, objects of a BankAccount class have private _name and _balance attributes that only the deposit() and withdraw() methods should directly access:

```
class BankAccount:
    def __init__(self, accountHolder):
        # BankAccount methods can access self._balance, but code outside of
        # this class should not:
❶       self._balance = 0
❷       self._name = accountHolder
        with open(self._name + 'Ledger.txt', 'w') as ledgerFile:
            ledgerFile.write('Balance is 0\n')

    def deposit(self, amount):
❸       if amount <= 0:
            return # Don't allow negative "deposits".
        self._balance += amount
❹       with open(self._name + 'Ledger.txt', 'a') as ledgerFile:
            ledgerFile.write('Deposit ' + str(amount) + '\n')
            ledgerFile.write('Balance is ' + str(self._balance) + '\n')

    def withdraw(self, amount):
❺       if self._balance < amount or amount < 0:
            return # Not enough in account, or withdraw is negative.
        self._balance -= amount
❻       with open(self._name + 'Ledger.txt', 'a') as ledgerFile:
            ledgerFile.write('Withdraw ' + str(amount) + '\n')
            ledgerFile.write('Balance is ' + str(self._balance) + '\n')

acct = BankAccount('Alice') # We create an account for Alice.
acct.deposit(120) # _balance can be affected through deposit()
acct.withdraw(40) # _balance can be affected through withdraw()

# Changing _name or _balance outside of BankAccount is impolite, but allowed:
❼ acct._balance = 1000000000
acct.withdraw(1000)

❽ acct._name = 'Bob' # Now we're modifying Bob's account ledger!
acct.withdraw(1000) # This withdrawal is recorded in BobLedger.txt!
```

When you run *privateExample.py*, the ledger files it creates are inaccurate because we modified the _balance and _name outside the class, which resulted in invalid states. *AliceLedger.txt* inexplicably has a lot of money in it:

```
Balance is 0
Deposit 120
Balance is 120
Withdraw 40
Balance is 80
Withdraw 1000
Balance is 999999000
```

Now there's a *BobLedger.txt* file with an inexplicable account balance, even though we never created a BankAccount object for Bob:

```
Withdraw 1000
Balance is 999998000
```

Well-designed classes will be mostly self-contained, providing methods to adjust the attributes to valid values. The _balance and _name attributes are marked as private ❶ ❷, and the only valid way of adjusting the BankAccount class's value is through the deposit() and withdraw() methods. These two methods have checks ❸ ❺ to make sure _balance isn't put into an invalid state (such as a negative integer value). These methods also record each transaction to account for the current balance ❹ ❻.

Code outside the class that modifies these attributes, such as acct._balance = 1000000000 ❼ or acct._name = 'Bob' ❽ instructions, can put the object into an invalid state and introduce bugs (and audits from the bank examiner). By following the underscore prefix convention for private access, you make debugging easier. The reason is that you know the cause of the bug will be in the code in the class instead of anywhere in the entire program.

Note that unlike Java and other languages, Python has no need for public getter and setter methods for private attributes. Instead Python uses properties, as explained in Chapter 17.

The type() Function and __qualname__ Attribute

Passing an object to the built-in type() function tells us the object's data type through its return value. The objects returned from the type() function are type objects, also called *class* objects. Recall that the terms *type*, *data type*, and *class* all have the same meaning in Python. To see what the type() function returns for various values, enter the following into the interactive shell:

```
>>> type(42)  # The object 42 has a type of int.
<class 'int'>
>>> int # int is a type object for the integer data type.
<class 'int'>
>>> type(42) == int  # Type check 42 to see if it is an integer.
```

```
True
>>> type('Hello') == int  # Type check 'Hello' against int.
False
>>> import wizcoin
>>> type(42) == wizcoin.WizCoin  # Type check 42 against WizCoin.
False
>>> purse = wizcoin.WizCoin(2, 5, 10)
>>> type(purse) == wizcoin.WizCoin # Type check purse against WizCoin.
True
```

Note that int is a type object and is the same kind of object as what type(42) returns, but it can also be called as the int() constructor function: the int('42') function doesn't convert the '42' string argument; instead, it returns an integer object based on the argument.

Say you need to log some information about the variables in your program to help you debug them later. You can only write strings to a logfile, but passing the type object to str() will return a rather messy-looking string. Instead, use the _qualname_ attribute, which all type objects have, to write a simpler, human-readable string:

```
>>> str(type(42))  # Passing the type object to str() returns a messy string.
"<class 'int'>"
>>> type(42).__qualname__ # The __qualname__ attribute is nicer looking.
'int'
```

The _qualname_ attribute is most often used for overriding the _repr_() method, which is explained in more detail in Chapter 17.

Non-OOP vs. OOP Examples: Tic-Tac-Toe

At first, it can be difficult to see how to use classes in your programs. Let's look at an example of a short tic-tac-toe program that doesn't use classes, and then rewrite it so it does.

Open a new file editor window and enter the following program; then save it as *tictactoe.py*:

```
# tictactoe.py, A non-OOP tic-tac-toe game.

ALL_SPACES = list('123456789')  # The keys for a TTT board dictionary.
X, O, BLANK = 'X', 'O', ' '  # Constants for string values.

def main():
    """Runs a game of tic-tac-toe."""
    print('Welcome to tic-tac-toe!')
    gameBoard = getBlankBoard()  # Create a TTT board dictionary.
    currentPlayer, nextPlayer = X, O  # X goes first, O goes next.

    while True:
        print(getBoardStr(gameBoard))  # Display the board on the screen.

        # Keep asking the player until they enter a number 1-9:
```

```
        move = None
        while not isValidSpace(gameBoard, move):
            print(f'What is {currentPlayer}\'s move? (1-9)')
            move = input()
        updateBoard(gameBoard, move, currentPlayer)  # Make the move.

        # Check if the game is over:
        if isWinner(gameBoard, currentPlayer):  # First check for victory.
            print(getBoardStr(gameBoard))
            print(currentPlayer + ' has won the game!')
            break
        elif isBoardFull(gameBoard):  # Next check for a tie.
            print(getBoardStr(gameBoard))
            print('The game is a tie!')
            break
        currentPlayer, nextPlayer = nextPlayer, currentPlayer  # Swap turns.
    print('Thanks for playing!')

def getBlankBoard():
    """Create a new, blank tic-tac-toe board."""
    board = {} # The board is represented as a Python dictionary.
    for space in ALL_SPACES:
        board[space] = BLANK  # All spaces start as blank.
    return board

def getBoardStr(board):
    """Return a text-representation of the board."""
    return f'''
      {board['1']}|{board['2']}|{board['3']}  1 2 3
      -+-+-
      {board['4']}|{board['5']}|{board['6']}  4 5 6
      -+-+-
      {board['7']}|{board['8']}|{board['9']}  7 8 9'''

def isValidSpace(board, space):
    """Returns True if the space on the board is a valid space number
    and the space is blank."""
    return space in ALL_SPACES or board[space] == BLANK

def isWinner(board, player):
    """Return True if player is a winner on this TTTBoard."""
    b, p = board, player # Shorter names as "syntactic sugar".
    # Check for 3 marks across the 3 rows, 3 columns, and 2 diagonals.
    return ((b['1'] == b['2'] == b['3'] == p) or # Across the top
            (b['4'] == b['5'] == b['6'] == p) or # Across the middle
            (b['7'] == b['8'] == b['9'] == p) or # Across the bottom
            (b['1'] == b['4'] == b['7'] == p) or # Down the left
            (b['2'] == b['5'] == b['8'] == p) or # Down the middle
            (b['3'] == b['6'] == b['9'] == p) or # Down the right
            (b['3'] == b['5'] == b['7'] == p) or # Diagonal
            (b['1'] == b['5'] == b['9'] == p))   # Diagonal

def isBoardFull(board):
    """Return True if every space on the board has been taken."""
```

```
    for space in ALL_SPACES:
        if board[space] == BLANK:
            return False  # If a single space is blank, return False.
    return True  # No spaces are blank, so return True.

def updateBoard(board, space, mark):
    """Sets the space on the board to mark."""
    board[space] = mark

if __name__ == '__main__':
    main() # Call main() if this module is run, but not when imported.
```

When you run this program, the output will look something like this:

```
Welcome to tic-tac-toe!

     | |   1 2 3
    -+-+-
     | |   4 5 6
    -+-+-
     | |   7 8 9
What is X's move? (1-9)
1

    X| |   1 2 3
    -+-+-
     | |   4 5 6
    -+-+-
     | |   7 8 9
What is O's move? (1-9)
--snip--
    X| |O  1 2 3
    -+-+- ,
     |O|   4 5 6
    -+-+-
    X|O|X  7 8 9
What is X's move? (1-9)
4

    X| |O  1 2 3
    -+-+-
    X|O|   4 5 6
    -+-+-
    X|O|X  7 8 9
X has won the game!
Thanks for playing!
```

Briefly, this program works by using a dictionary object to represent the nine spaces on a tic-tac-toe board. The dictionary's keys are the strings '1' through '9', and its values are the strings 'X', 'O', or ' '. The numbered spaces are in the same arrangement as a phone's keypad.

The functions in *tictactoe.py* do the following:

- The main() function contains the code that creates a new board data structure (stored in the gameBoard variable) and calls other functions in the program.
- The getBlankBoard() function returns a dictionary with the nine spaces set to ' ' for a blank board.
- The getBoardStr() function accepts a dictionary representing the board and returns a multiline string representation of the board that can be printed to the screen. This is what renders the tic-tac-toe board's text that the game displays.
- The isValidSpace() function returns True if it's passed a valid space number and that space is blank.
- The isWinner() function's parameters accept a board dictionary and either 'X' or 'O' to determine whether that player has three marks in a row on the board.
- The isBoardFull() function determines whether the board has no blank spaces, meaning the game has ended. The updateBoard() function's parameters accept a board dictionary, a space, and a player's X or O mark and updates the dictionary.

Notice that many of the functions accept the variable board as their first parameter. That means these functions are related to each other in that they all operate on a common data structure.

When several functions in the code all operate on the same data structure, it's usually best to group them together as the methods and attributes of a class. Let's redesign this in the *tictactoe.py* program to use a TTTBoard class that will store the board dictionary in an attribute named spaces. The functions that had board as a parameter will become methods of our TTTBoard class and use the self parameter instead of a board parameter.

Open a new file editor window, enter the following code, and save it as *tictactoe_oop.py*:

```
# tictactoe_oop.py, an object-oriented tic-tac-toe game.

ALL_SPACES = list('123456789')  # The keys for a TTT board.
X, O, BLANK = 'X', 'O', ' '  # Constants for string values.

def main():
    """Runs a game of tic-tac-toe."""
    print('Welcome to tic-tac-toe!')
    gameBoard = TTTBoard()  # Create a TTT board object.
    currentPlayer, nextPlayer = X, O # X goes first, O goes next.

    while True:
        print(gameBoard.getBoardStr())  # Display the board on the screen.

        # Keep asking the player until they enter a number 1-9:
        move = None
```

```python
        while not gameBoard.isValidSpace(move):
            print(f'What is {currentPlayer}\'s move? (1-9)')
            move = input()
        gameBoard.updateBoard(move, currentPlayer)  # Make the move.

        # Check if the game is over:
        if gameBoard.isWinner(currentPlayer):  # First check for victory.
            print(gameBoard.getBoardStr())
            print(currentPlayer + ' has won the game!')
            break
        elif gameBoard.isBoardFull():  # Next check for a tie.
            print(gameBoard.getBoardStr())
            print('The game is a tie!')
            break
        currentPlayer, nextPlayer = nextPlayer, currentPlayer  # Swap turns.
    print('Thanks for playing!')

class TTTBoard:
    def __init__(self, usePrettyBoard=False, useLogging=False):
        """Create a new, blank tic tac toe board."""
        self._spaces = {}  # The board is represented as a Python dictionary.
        for space in ALL_SPACES:
            self._spaces[space] = BLANK  # All spaces start as blank.

    def getBoardStr(self):
        """Return a text-representation of the board."""
        return f'''
  {self._spaces['1']}|{self._spaces['2']}|{self._spaces['3']}  1 2 3
  -+-+-
  {self._spaces['4']}|{self._spaces['5']}|{self._spaces['6']}  4 5 6
  -+-+-
  {self._spaces['7']}|{self._spaces['8']}|{self._spaces['9']}  7 8 9'''

    def isValidSpace(self, space):
        """Returns True if the space on the board is a valid space number
        and the space is blank."""
        return space in ALL_SPACES and self._spaces[space] == BLANK

    def isWinner(self, player):
        """Return True if player is a winner on this TTTBoard."""
        s, p = self._spaces, player # Shorter names as "syntactic sugar".
        # Check for 3 marks across the 3 rows, 3 columns, and 2 diagonals.
        return ((s['1'] == s['2'] == s['3'] == p) or # Across the top
                (s['4'] == s['5'] == s['6'] == p) or # Across the middle
                (s['7'] == s['8'] == s['9'] == p) or # Across the bottom
                (s['1'] == s['4'] == s['7'] == p) or # Down the left
                (s['2'] == s['5'] == s['8'] == p) or # Down the middle
                (s['3'] == s['6'] == s['9'] == p) or # Down the right
                (s['3'] == s['5'] == s['7'] == p) or # Diagonal
                (s['1'] == s['5'] == s['9'] == p))   # Diagonal

    def isBoardFull(self):
        """Return True if every space on the board has been taken."""
```

```
        for space in ALL_SPACES:
            if self._spaces[space] == BLANK:
                return False  # If a single space is blank, return False.
        return True  # No spaces are blank, so return True.

    def updateBoard(self, space, player):
        """Sets the space on the board to player."""
        self._spaces[space] = player

if __name__ == '__main__':
    main() # Call main() if this module is run, but not when imported.
```

Functionally, this program is the same as the non-OOP tic-tac-toe program. The output looks identical. We've moved the code that used to be in getBlankBoard() to the TTTBoard class's __init__() method, because they perform the same task of preparing the board data structure. We converted the other functions into methods, with the self parameter replacing the old board parameter, because they also serve a similar purpose: they're both blocks of code that operate on a tic-tac-toe board data structure.

When the code in these methods needs to change the dictionary stored in the _spaces attribute, the code uses self._spaces. When the code in these methods need to call other methods, the calls would also be preceded by self and a period. This is similar to how coinJars.values() in "Creating a Simple Class: WizCoin" had an object in the coinJars variable. In this example, the object that has the method to call is in a self variable.

Also, notice that the _spaces attribute begins with an underscore, meaning that only code inside the methods of TTTBoard should access or modify it. Code outside the class should only be able to modify _spaces indirectly by calling methods that modify it.

It can be helpful to compare the source code of the two tic-tac-toe programs. You can compare the code in this book or view a side-by-side comparison at *https://autbor.com/compareoop/*.

Tic-tac-toe is a small program, so it doesn't take much effort to understand. But what if this program were tens of thousands of lines long with hundreds of different functions? A program with a few dozen classes would be easier to understand than a program with several hundred disparate functions. OOP breaks down a complicated program into easier-to-understand chunks.

Designing Classes for the Real World Is Hard

Designing a class, just like designing a paper form, seems deceptively straightforward. Forms and classes are, by their nature, simplifications of the real-world objects they represent. The question is, how should we simplify these objects? For example, if we're creating a Customer class, the customer should have a firstName and lastName attribute, right? But actually creating classes to model real-world objects can be tricky. In most Western countries, a person's last name is their family name, but in China, the family name is first. If we don't want to exclude more than one billion potential customers, how should we change our `Customer` class? Should we change `firstName` and `lastName` to `givenName` and `familyName`? But some cultures don't use family names. For example, former UN Secretary General U Thant, who is Burmese, has no family name: Thant is his given name and U is an initialization of his father's given name. We might want to record the customer's age, but an age attribute would soon become out of date; instead, it's best to calculate the age each time you need it using a `birthdate` attribute.

The real world is complicated, and designing forms and classes to capture this complexity in a uniform structure on which our programs can operate is difficult. Phone number formats vary between countries. ZIP codes don't apply to addresses outside the United States. Setting a maximum number of characters for city names could be a problem for the German hamlet of Schmedeswurtherwesterdeich. In Australia and New Zealand, your legally recognized gender can be X. A platypus is a mammal that lays eggs. A peanut is not a nut. A hotdog might or might not be a sandwich, depending on who you ask. As a programmer writing programs for use in the real world, you'll have to navigate this complexity.

To learn more about this topic, I recommend the PyCon 2015 talk "Schemas for the Real World" by Carina C. Zona at *https://youtu.be/PYYfVqtcWQY/* and the North Bay Python 2018 talk "Hi! My name is . . ." by James Bennett at *https://youtu.be/NIebellpdYk/*. There are also popular "Falsehoods Programmers Believe" blog posts, such as "Falsehoods Programmers Believe About Names" and "Falsehoods Programmers Believe About Time Zones." These blog posts also cover topics like maps, email addresses, and many more kinds of data that programmers often poorly represent. You'll find a collection of links to these articles at *https://github.com/kdeldycke/awesome-falsehood/*. Additionally, you'll find a good example of a poorly executed method of capturing real-world complexity in CGP Grey's video, "Social Security Cards Explained," at *https://youtu.be/Erp8IAUouus/*.

Summary

OOP is a useful feature for organizing your code. Classes allow you to group together data and code into new data types. You can also create objects from these classes by calling their constructors (the class's name called as a function), which in turn, calls the class's __init__() method. Methods are functions associated with objects, and attributes are variables associated with objects. All methods have a `self` parameter as their first parameter, which is assigned the object when the method is called. This allows the methods to read or set the object's attributes and call its methods.

Although Python doesn't allow you to specify private or public access for attributes, it does have a convention of using an underscore prefix for any methods or attributes that should only be called or accessed from the class's own methods. By following this convention, you can avoid misusing the class and setting it into an invalid state that could cause bugs. Calling type(obj) will return the obj type's class object. Class objects have a __qualname__ attribute, which contains a string with a human-readable form of the class's name.

At this point, you might be thinking, why we should bother using classes, attributes, and methods when we could do the same task with functions? OOP is a useful way to organize your code into more than just a *.py* file with 100 functions in it. By breaking up your program into several well-designed classes, you can focus on each class separately.

OOP is an approach that focuses on data structures and the methods to handle those data structures. This approach isn't mandatory for every program, and it's certainly possible to overuse OOP. But OOP provides opportunities to use many advanced features that we'll explore in the next two chapters. The first of these features is inheritance, which we'll delve into in the next chapter.

16

OBJECT-ORIENTED PROGRAMMING AND INHERITANCE

Defining a function and calling it from several places saves you from having to copy and paste source code. Not duplicating code is a good practice, because if you need to change it (either for a bug fix or to add new features), you only need to change it in one place. Without duplicate code, the program is also shorter and easier to read.

Similar to functions, *inheritance* is a code reuse technique that you can apply to classes. It's the act of putting classes into parent-child relationships in which the child class inherits a copy of the parent class's methods, freeing you from duplicating a method in multiple classes.

Many programmers think inheritance is overrated or even dangerous because of the added complexity that large webs of inherited classes add to a program. Blog posts with titles like "Inheritance Is Evil" are not entirely off the mark; inheritance is certainly easy to overuse. But limited use of this technique can be a huge time-saver when it comes to organizing your code.

How Inheritance Works

To create a new child class, you put the name of the existing parent class in between parentheses in the class statement. To practice creating a child class, open a new file editor window and enter the following code; save it as *inheritanceExample.py*:

```
❶ class ParentClass:
❷     def printHello(self):
           print('Hello, world!')

❸ class ChildClass(ParentClass):
       def someNewMethod(self):
           print('ParentClass objects don\'t have this method.')

❹ class GrandchildClass(ChildClass):
       def anotherNewMethod(self):
           print('Only GrandchildClass objects have this method.')

print('Create a ParentClass object and call its methods:')
parent = ParentClass()
parent.printHello()

print('Create a ChildClass object and call its methods:')
child = ChildClass()
child.printHello()
child.someNewMethod()

print('Create a GrandchildClass object and call its methods:')
grandchild = GrandchildClass()
grandchild.printHello()
grandchild.someNewMethod()
grandchild.anotherNewMethod()

print('An error:')
parent.someNewMethod()
```

When you run this program, the output should look like this:

```
Create a ParentClass object and call its methods:
Hello, world!
Create a ChildClass object and call its methods:
Hello, world!
ParentClass objects don't have this method.
Create a GrandchildClass object and call its methods:
Hello, world!
ParentClass objects don't have this method.
Only GrandchildClass objects have this method.
An error:
Traceback (most recent call last):
  File "inheritanceExample.py", line 35, in <module>
    parent.someNewMethod() # ParentClass objects don't have this method.
AttributeError: 'ParentClass' object has no attribute 'someNewMethod'
```

We've created three classes named ParentClass ❶, ChildClass ❸, and GrandchildClass ❹. The ChildClass *subclasses* ParentClass, meaning that ChildClass will have all the same methods as ParentClass. We say that ChildClass *inherits* methods from ParentClass. Also, GrandchildClass subclasses ChildClass, so it has all the same methods as ChildClass and its parent, ParentClass.

Using this technique, we've effectively copied and pasted the code for the printHello() method ❷ into the ChildClass and GrandchildClass classes. Any changes we make to the code in printHello() update not only ParentClass, but also ChildClass and GrandchildClass. This is the same as changing the code in a function updates all of its function calls. You can see this relationship in Figure 16-1. Notice that in class diagrams, the arrow is drawn from the subclass pointing to the base class. This reflects the fact that a class will always know its base class but won't know its subclasses.

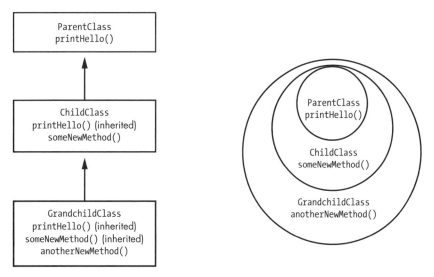

Figure 16-1: A hierarchical diagram (left) and Venn diagram (right) showing the relationships between the three classes and the methods they have

It's common to say that parent-child classes represent "is a" relationships. A ChildClass object is a ParentClass object because it has all the same methods that a ParentClass object has, including some additional methods it defines. This relationship is one way: a ParentClass object is not a ChildClass object. If a ParentClass object tries to call someNewMethod(), which only exists for ChildClass objects (and the subclasses of ChildClass), Python raises an AttributeError.

Programmers often think of related classes as having to fit into some real-world "is a" hierarchy. OOP tutorials commonly have parent, child, and grandchild classes of Vehicle ▸ FourWheelVehicle ▸ Car, Animal ▸ Bird ▸ Sparrow, or Shape ▸ Rectangle ▸ Square. But remember that the primary purpose of inheritance is code reuse. If your program needs a class with a set of methods that is a complete superset of some other class's methods, inheritance allows you to avoid copying and pasting code.

We also sometimes call a child class a *subclass* or *derived class* and call a parent class the *super class* or *base class*.

Overriding Methods

Child classes inherit all the methods of their parent classes. But a child class can override an inherited method by providing its own method with its own code. The child class's overriding method will have the same name as the parent class's method.

To illustrate this concept, let's return to the tic-tac-toe game we created in the previous chapter. This time, we'll create a new class, MiniBoard, that subclasses TTTBoard and overrides getBoardStr() to provide a smaller drawing of the tic-tac-toe board. The program will ask the player which board style to use. We don't need to copy and paste the rest of the TTTBoard methods because MiniBoard will inherit them.

Add the following to the end of your *tictactoe_oop.py* file to create a child class of the original TTTBoard class and then override the getBoardStr() method:

```
class MiniBoard(TTTBoard):
    def getBoardStr(self):
        """Return a tiny text-representation of the board."""
        # Change blank spaces to a '.'
        for space in ALL_SPACES:
            if self._spaces[space] == BLANK:
                self._spaces[space] = '.'

        boardStr = f'''
    {self._spaces['1']}{self._spaces['2']}{self._spaces['3']} 123
    {self._spaces['4']}{self._spaces['5']}{self._spaces['6']} 456
    {self._spaces['7']}{self._spaces['8']}{self._spaces['9']} 789'''

        # Change '.' back to blank spaces.
        for space in ALL_SPACES:
            if self._spaces[space] == '.':
                self._spaces[space] = BLANK
        return boardStr
```

As with the getBoardStr() method for the TTTBoard class, the getBoardStr() method for MiniBoard creates a multiline string of a tic-tac-toe board to display when passed to the print() function. But this string is much smaller, forgoing the lines between the X and O marks and using periods to indicate blank spaces.

Change the line in main() so it instantiates a MiniBoard object instead of a TTTBoard object:

```
if input('Use mini board? Y/N: ').lower().startswith('y'):
    gameBoard = MiniBoard() # Create a MiniBoard object.
else:
    gameBoard = TTTBoard() # Create a TTTBoard object.
```

Other than this one line change to `main()`, the rest of the program works the same as before. When you run the program now, the output will look something like this:

```
Welcome to Tic-Tac-Toe!
Use mini board? Y/N: y

          ... 123
          ... 456
          ... 789
What is X's move? (1-9)
1

          X.. 123
          ... 456
          ... 789
What is O's move? (1-9)
--snip--
          XXX 123
          .OO 456
          O.X 789
X has won the game!
Thanks for playing!
```

Your program can now easily have both implementations of these tic-tac-toe board classes. Of course, if you *only* want the mini version of the board, you could simply replace the code in the getBoardStr() method for TTTBoard. But if you need *both*, inheritance lets you easily create two classes by reusing their common code.

If we didn't use inheritance, we could have, say, added a new attribute to TTTBoard called useMiniBoard and put an if-else statement inside getBoardStr() to decide when to show the regular board or the mini one. This would work well for such a simple change. But what if the MiniBoard subclass needed to override 2, 3, or even 100 methods? What if we wanted to create several different subclasses of TTTBoard? Not using inheritance would cause an explosion of if-else statements inside our methods and a large increase in our code's complexity. By using subclasses and overriding methods, we can better organize our code into separate classes to handle these different use cases.

The super() Function

A child class's overridden method is often similar to the parent class's method. Even though inheritance is a code reuse technique, overriding a method might cause you to rewrite the same code from the parent class's method as part of the child class's method. To prevent this duplicate code, the built-in super() function allows an overriding method to call the original method in the parent class.

For example, let's create a new class called HintBoard that subclasses TTTBoard. The new class overrides getBoardStr(), so after drawing the tic-tac-toe board, it also adds a hint if either X or O could win on their next move. This means that the HintBoard class's getBoardStr() method has to do all the same tasks that the TTTBoard class's getBoardStr() method does to draw the tic-tac-toe board. Instead of repeating the code to do this, we can use super() to call the TTTBoard class's getBoardStr() method from the HintBoard class's getBoardStr() method. Add the following to the end of your *tictactoe _oop.py* file:

```
class HintBoard(TTTBoard):
    def getBoardStr(self):
        """Return a text-representation of the board with hints."""
❶       boardStr = super().getBoardStr() # Call getBoardStr() in TTTBoard.

        xCanWin = False
        oCanWin = False
❷       originalSpaces = self._spaces # Backup _spaces.
        for space in ALL_SPACES: # Check each space:
            # Simulate X moving on this space:
            self._spaces = copy.copy(originalSpaces)
            if self._spaces[space] == BLANK:
                self._spaces[space] = X
            if self.isWinner(X):
                xCanWin = True
            # Simulate O moving on this space:
❸           self._spaces = copy.copy(originalSpaces)
            if self._spaces[space] == BLANK:
                self._spaces[space] = O
            if self.isWinner(O):
                oCanWin = True
        if xCanWin:
            boardStr += '\nX can win in one more move.'
        if oCanWin:
            boardStr += '\nO can win in one more move.'
        self._spaces = originalSpaces
        return boardStr
```

First, super().getBoardStr() ❶ runs the code inside the parent TTTBoard class's getBoardStr(), which returns a string of the tic-tac-toe board. We save this string in a variable named boardStr for now. With the board string created by reusing TTTBoard class's getBoardStr(), the rest of the code in this method handles generating the hint. The getBoardStr() method then sets xCanWin and oCanWin variables to False, and backs up the self._spaces dictionary to an originalSpaces variable ❷. Then a for loop loops over all board spaces from '1' to '9'. Inside the loop, the self._spaces attribute is set to a copy of the originalSpaces dictionary, and if the current space being looped on is blank, an X is placed there. This simulates X moving on this blank space for its next move. A call to self.isWinner() will determine if this would be a winning move, and if so, xCanWin is set to True. Then these steps are repeated for O to see whether O could win by moving on this space ❸. This method uses the copy module to make a copy of the dictionary in self._spaces, so add the following line to the top of *tictactoe.py*:

```
import copy
```

Next, change the line in main() so it instantiates a HintBoard object instead of a TTTBoard object:

```
gameBoard = HintBoard() # Create a TTT board object.
```

Other than this one line change to main(), the rest of the program works exactly as before. When you run the program now, the output will look something like this:

```
Welcome to Tic-Tac-Toe!
--snip--
    X| |   1 2 3
    -+-+-
    | |0   4 5 6
    -+-+-
    | |X   7 8 9
X can win in one more move.
What is 0's move? (1-9)
5

    X| |   1 2 3
    -+-+-
    |0|0   4 5 6
    -+-+-
    | |X   7 8 9
0 can win in one more move.
--snip--
The game is a tie!
Thanks for playing!
```

At the end of the method, if xCanWin or oCanWin is True, an additional message stating so is added to the boardStr string. Finally, boardStr is returned.

Not every overridden method needs to use super()! If a class's overriding method does something completely different from the overridden method in the parent class, there's no need to call the overridden method using super(). The super() function is especially useful when a class has more than one parent method, as explained in "Multiple Inheritance" later in this chapter.

Favor Composition Over Inheritance

Inheritance is a great technique for code reuse, and you might want to start using it immediately in all your classes. But you might not always want the base and subclasses to be so tightly coupled. Creating multiple levels of inheritance doesn't add organization so much as bureaucracy to your code.

Although you can use inheritance for classes with "is a" relationships (in other words, when the child class is a kind of the parent class), it's often favorable to use a technique called *composition* for classes with "has a" relationships. Composition is the class design technique of including objects in your class rather than inheriting those objects' class. This is what we do when we add attributes to our classes. When designing your classes using inheritance, favor composition instead of inheritance. This is what we've been doing with all the examples in this and the previous chapter, as described in the following list:

- A WizCoin object "has an" amount of galleon, sickle, and knut coins.
- A TTTBoard object "has a" set of nine spaces.
- A MiniBoard object "is a" TTTBoard object, so it also "has a" set of nine spaces.
- A HintBoard object "is a" TTTBoard object, so it also "has a" set of nine spaces.

Let's return to our WizCoin class from the previous chapter. If we created a new WizardCustomer class to represent customers in the wizarding world, those customers would be carrying an amount of money, which we could represent through the WizCoin class. But there is no "is a" relationship between the two classes; a WizardCustomer object is not a kind of WizCoin object. If we used inheritance, it could create some awkward code:

```
import wizcoin

❶ class WizardCustomer(wizcoin.WizCoin):
    def __init__(self, name):
        self.name = name
        super().__init__(0, 0, 0)

wizard = WizardCustomer('Alice')
print(f'{wizard.name} has {wizard.value()} knuts worth of money.')
print(f'{wizard.name}\'s coins weigh {wizard.weightInGrams()} grams.')
```

In this example, WizardCustomer inherits the methods of a WizCoin ❶ object, such as value() and weightInGrams(). Technically, a WizardCustomer that inherits from WizCoin can do all the same tasks that a WizardCustomer that includes a WizCoin object as an attribute can. But the wizard.value() and wizard.weightInGrams() method names are misleading: it seems like they would return the wizard's value and weight rather than the value and weight of the wizard's coins. In addition, if we later wanted to add a weightInGrams() method for the wizard's weight, that method name would already be taken.

It's much better to have a WizCoin object as an attribute, because a wizard customer "has a" quantity of wizard coins:

```
import wizcoin

class WizardCustomer:
    def __init__(self, name):
        self.name = name
❶       self.purse = wizcoin.WizCoin(0, 0, 0)

wizard = WizardCustomer('Alice')
print(f'{wizard.name} has {wizard.purse.value()} knuts worth of money.')
print(f'{wizard.name}\'s coins weigh {wizard.purse.weightInGrams()} grams.')
```

Instead of making the `WizardCustomer` class inherit methods from `WizCoin`, we give the `WizardCustomer` class a purse attribute ❶, which contains a `WizCoin` object. When you use composition, any changes to the `WizCoin` class's methods won't change the `WizardCustomer` class's methods. This technique offers more flexibility in future design changes for both classes and leads to more maintainable code.

Inheritance's Downside

The primary downside of inheritance is that any future changes you make to parent classes are necessarily inherited by all its child classes. In most cases, this tight coupling is exactly what you want. But in some instances, your code requirements won't easily fit your inheritance model.

For example, let's say we have `Car`, `Motorcycle`, and `LunarRover` classes in a vehicle simulation program. They all need similar methods, such as `startIgnition()` and `changeTire()`. Instead of copying and pasting this code into each class, we can create a parent `Vehicle` class and have `Car`, `Motorcycle`, and `LunarRover` inherit it. Now if we need to fix a bug in, say, the `changeTire()` method, there's only one place we need to make the change. This is especially helpful if we have dozens of different vehicle-related classes inheriting from `Vehicle`. The code for these classes would look like this:

```
class Vehicle:
    def __init__(self):
        print('Vehicle created.')
    def startIgnition(self):
        pass  # Ignition starting code goes here.
    def changeTire(self):
        pass  # Tire changing code goes here.

class Car(Vehicle):
    def __init__(self):
        print('Car created.')

class Motorcycle(Vehicle):
    def __init__(self):
        print('Motorcycle created.')

class LunarRover(Vehicle):
    def __init__(self):
        print('LunarRover created.')
```

But all future changes to `Vehicle` will affect these subclasses as well. What happens if we need a `changeSparkPlug()` method? Cars and motorcycles have combustion engines with spark plugs, but lunar rovers don't. By favoring composition over inheritance, we can create separate `CombustionEngine` and `ElectricEngine` classes. Then we design the `Vehicle` class so it "has an" engine attribute, either a `CombustionEngine` or `ElectricEngine` object, with the appropriate methods:

```
class CombustionEngine:
    def __init__(self):
        print('Combustion engine created.')
    def changeSparkPlug(self):
        pass  # Spark plug changing code goes here.

class ElectricEngine:
    def __init__(self):
        print('Electric engine created.')

class Vehicle:
    def __init__(self):
        print('Vehicle created.')
        self.engine = CombustionEngine()  # Use this engine by default.
--snip--

class LunarRover(Vehicle):
    def __init__(self):
        print('LunarRover created.')
        self.engine = ElectricEngine()
```

This could require rewriting large amounts of code, particularly if you have several classes that inherit from a preexisting `Vehicle` class: all the `vehicleObj.changeSparkPlug()` calls would need to become `vehicleObj.engine .changeSparkPlug()` for every object of the `Vehicle` class or its subclasses. Because such a sizeable change could introduce bugs, you might want to simply have the `changeSparkPlug()` method for `LunarVehicle` do nothing. In this case, the Pythonic way is to set `changeSparkPlug` to `None` inside the `LunarVehicle` class:

```
class LunarRover(Vehicle):
    changeSparkPlug = None
    def __init__(self):
        print('LunarRover created.')
```

The `changeSparkPlug = None` line follows the syntax described in "Class Attributes" later in this chapter. This overrides the `changeSparkPlug()` method inherited from `Vehicle`, so calling it with a `LunarRover` object causes an error:

```
>>> myVehicle = LunarRover()
LunarRover created.
>>> myVehicle.changeSparkPlug()
Traceback (most recent call last):
  File "<stdin>", line 1, in <module>
TypeError: 'NoneType' object is not callable
```

This error allows us to fail fast and immediately see a problem if we try to call this inappropriate method with a `LunarRover` object. Any child classes of `LunarRover` also inherit this `None` value for `changeSparkPlug()`. The `TypeError: 'NoneType' object is not callable` error message tells us that the programmer of the LunarRover class intentionally set the changeSparkPlug() method to None. If no such method existed in the first place, we would have received a NameError: name 'changeSparkPlug' is not defined error message.

Inheritance can create classes with complexity and contradiction. It's often favorable to use composition instead.

The isinstance() and issubclass() Functions

When we need to know the type of an object, we can pass the object to the built-in type() function, as described in the previous chapter. But if we're doing a type check of an object, it's a better idea to use the more flexible isinstance() built-in function. The isinstance() function will return True if the object is of the given class *or a subclass of the given class*. Enter the following into the interactive shell:

```
>>> class ParentClass:
...     pass
...
>>> class ChildClass(ParentClass):
...     pass
...
>>> parent = ParentClass() # Create a ParentClass object.
>>> child = ChildClass() # Create a ChildClass object.
>>> isinstance(parent, ParentClass)
True
>>> isinstance(parent, ChildClass)
False
❶ >>> isinstance(child, ChildClass)
True
❷ >>> isinstance(child, ParentClass)
True
```

Notice that isinstance() indicates that the ChildClass object in child is an instance of ChildClass ❶ and an instance of ParentClass ❷. This makes sense, because a ChildClass object "is a" kind of ParentClass object.

You can also pass a tuple of class objects as the second argument to see whether the first argument is one of any of the classes in the tuple:

```
>>> isinstance(42, (int, str, bool)) # True if 42 is an int, str, or bool.
True
```

The less commonly used issubclass() built-in function can identify whether the class object passed for the first argument is a subclass of (or the same class as) the class object passed for the second argument:

```
>>> issubclass(ChildClass, ParentClass) # ChildClass subclasses ParentClass.
True
>>> issubclass(ChildClass, str) # ChildClass doesn't subclass str.
False
>>> issubclass(ChildClass, ChildClass) # ChildClass is ChildClass.
True
```

As you can with isinstance(), you can pass a tuple of class objects as the second argument to issubclass() to see whether the first argument is a subclass of any of the classes in the tuple. The key difference between isinstance() and issubclass() is that issubclass() is passed two class objects, whereas isinstance() is passed an object and a class object.

Class Methods

Class methods are associated with a class rather than with individual objects, like regular methods are. You can recognize a class method in code when you see two markers: the @classmethod decorator before the method's def statement and the use of cls as the first parameter, as shown in the following example.

```
class ExampleClass:
    def exampleRegularMethod(self):
        print('This is a regular method.')

    @classmethod
    def exampleClassMethod(cls):
        print('This is a class method.')

# Call the class method without instantiating an object:
ExampleClass.exampleClassMethod()

obj = ExampleClass()
# Given the above line, these two lines are equivalent:
obj.exampleClassMethod()
obj.__class__.exampleClassMethod()
```

The cls parameter acts like self except self refers to an object, but the cls parameter refers to an object's class. This means that the code in a class method cannot access an individual object's attributes or call an object's regular methods. Class methods can only call other class methods or access class attributes. We use the name cls because class is a Python keyword, and just like other keywords, such as if, while, or import, we can't use it for parameter names. We often call class attributes through the class object, as in ExampleClass.exampleClassMethod(). But we can also call them through any object of the class, as in obj.exampleClassMethod().

Class methods aren't commonly used. The most frequent use case is to provide alternative constructor methods besides __init__(). For example, what if a constructor function could accept either a string of data the new object needs or a string of a filename that contains the data the new object needs? We don't want the list of the __init__() method's parameters to be lengthy and confusing. Instead let's use class methods to return a new object.

For example, let's create an AsciiArt class. As you saw in Chapter 14, ASCII art uses text characters to form an image.

```
class AsciiArt:
    def __init__(self, characters):
        self._characters = characters

    @classmethod
    def fromFile(cls, filename):
        with open(filename) as fileObj:
            characters = fileObj.read()
            return cls(characters)

    def display(self):
        print(self._characters)

    # Other AsciiArt methods would go here...

face1 = AsciiArt(' _____\n' +
                 '|  . .  |\n' +
                 '| \\\\__/ |\n' +
                 '|_____|')
face1.display()

face2 = AsciiArt.fromFile('face.txt')
face2.display()
```

The AsciiArt class has an __init__() method that can be passed the text characters of the image as a string. It also has a fromFile() class method that can be passed the filename string of a text file containing the ASCII art. Both methods create AsciiArt objects.

When you run this program and there is a *face.txt* file that contains the ASCII art face, the output will look something like this:

The fromFile() class method makes your code a bit easier to read, compared to having __init__() do everything.

Another benefit of class methods is that a subclass of AsciiArt can inherit its fromFile() method (and override it if necessary). This is why we call cls(characters) in the AsciiArt class's fromFile() method instead of AsciiArt(characters). The cls() call will also work in subclasses of AsciiArt without modification because the AsciiArt class isn't hardcoded into the method. But an AsciiArt() call would always call AsciiArt class's __init__() instead of the subclass's __init__(). You can think of cls as meaning "an object representing this class."

Keep in mind that just as regular methods should always use their self parameter somewhere in their code, a class method should always use its cls parameter. If your class method's code *never* uses the cls parameter, it's a sign that your class method should probably just be a function.

Class Attributes

A class attribute is a variable that belongs to the class rather than to an object. We create class attributes inside the class but outside all methods, just like we create global variables in a *.py* file but outside all functions. Here's an example of a class attribute named count, which keeps track of how many CreateCounter objects have been created:

```
class CreateCounter:
    count = 0 # This is a class attribute.

    def __init__(self):
        CreateCounter.count += 1

print('Objects created:', CreateCounter.count)  # Prints 0.
a = CreateCounter()
b = CreateCounter()
c = CreateCounter()
print('Objects created:', CreateCounter.count)  # Prints 3.
```

The CreateCounter class has a single class attribute named count. All CreateCounter objects share this attribute rather than having their own separate count attributes. This is why the CreateCounter.count += 1 line in the constructor function can keep count of every CreateCounter object created. When you run this program, the output will look like this:

```
Objects created: 0
Objects created: 3
```

We rarely use class attributes. Even this "count how many CreateCounter objects have been created" example can be done more simply by using a global variable instead of a class attribute.

Static Methods

A *static method* doesn't have a self or cls parameter. Static methods are effectively just functions, because they can't access the attributes or methods of the class or its objects. Rarely, if ever, do you need to use static methods in Python. If you do decide to use one, you should strongly consider just creating a regular function instead.

We define static methods by placing the @staticmethod decorator before their def statements. Here is an example of a static method.

```
class ExampleClassWithStaticMethod:
    @staticmethod
    def sayHello():
        print('Hello!')

# Note that no object is created, the class name precedes sayHello():
ExampleClassWithStaticMethod.sayHello()
```

There would be almost no difference between the sayHello() static method in the ExampleClassWithStaticMethod class and a sayHello() function. In fact, you might prefer to use a function, because you can call it without having to enter the class name beforehand.

Static methods are more common in other languages that don't have Python's flexible language features. Python's inclusion of static methods imitates the features of other languages but doesn't offer much practical value.

When to Use Class and Static Object-Oriented Features

You'll rarely need class methods, class attributes, and static methods. They're also prone to overuse. If you're thinking, "Why can't I just use a function or global variable instead?" this is a hint that you probably don't need to use a class method, class attribute, or static method. The only reason this intermediate-level book covers them is so you can recognize them when you encounter them in code, but I'm not encouraging you to use them. They can be useful if you're creating your own framework with an elaborate family of classes that are, in turn, expected to be subclassed by programmers using the framework. But you most likely won't need them when you're writing straightforward Python applications.

For more discussion on these features and why you do or don't need them, read Phillip J. Eby's post "Python Is Not Java" at *https://dirtsimple.org/2004/12/python-is-not-java.html* and Ryan Tomayko's "The Static Method Thing" at *https://tomayko.com/blog/2004/the-static-method-thing*.

Object-Oriented Buzzwords

Explanations of OOP often begin with a lot of jargon, such as inheritance, encapsulation, and polymorphism. The importance of knowing these terms is overrated, but you should have at least a basic understanding of them. I already covered inheritance, so I'll describe the other concepts here.

Encapsulation

The word *encapsulation* has two common but related definitions. The first definition is that encapsulation is the bundling of related data and code into a single unit. To encapsulate means to *box up*. This is essentially what classes do: they combine related attributes and methods. For example, our WizCoin class encapsulates three integers for knuts, sickles, and galleons into a single WizCoin object.

The second definition is that encapsulation is an *information hiding* technique that lets objects hide complex implementation details about how the object works. You saw this in "Private Attributes and Private Methods" on page 282, where BankAccount objects present deposit() and withdraw() methods to hide the details of how their _balance attributes are handled. Functions serve a similar *black box* purpose: how the math.sqrt() function calculates the square root of a number is hidden. All you need to know is that the function returns the square root of the number you passed it.

Polymorphism

Polymorphism allows objects of one type to be treated as objects of another type. For example, the len() function returns the length of the argument passed to it. You can pass a string to len() to see how many characters it has, but you can also pass a list or dictionary to len() to see how many items or key-value pairs it has, respectively. This form of polymorphism is called *generic functions* or *parametric polymorphism*, because it can handle objects of many different types.

Polymorphism also refers to *ad hoc polymorphism* or *operator overloading*, where operators (such as + or *) can have different behavior based on the type of objects they're operating on. For example, the + operator does mathematical addition when operating on two integer or float values, but it does string concatenation when operating on two strings. Operator overloading is covered in Chapter 17.

When Not to Use Inheritance

It's easy to overengineer your classes using inheritance. As Luciano Ramalho states, "Placing objects in a neat hierarchy appeals to our sense of order; programmers do it just for fun." We'll create classes, subclasses, and sub-subclasses when a single class, or a couple of functions in a module, would achieve the same effect. But recall the Zen of Python tenet in Chapter 6 that *simple is better than complex.*

Using OOP allows you to organize your code into smaller units (in this case, classes) that are easier to reason about than one large *.py* file with hundreds of functions defined in no particular order. Inheritance is useful if you have several functions that all operate on the same dictionary or list data structure. In that case, it's beneficial to organize them into a class.

But here are some examples of when you don't need to create a class or use inheritance:

- If your class consists of methods that never use the self or cls parameter, delete the class and use functions in place of the methods.

- If you've created a parent with only a single child class but never create objects of the parent class, you can combine them into a single class.

- If you create more than three or four levels of subclasses, you're probably using inheritance unnecessarily. Combine those subclasses into fewer classes.

As the non-OOP and OOP versions of the tic-tac-toe program in the previous chapter illustrate, it's certainly possible to not use classes and still have a working, bug-free program. Don't feel that you have to design your program as some complex web of classes. A simple solution that works is better than a complicated solution that doesn't. Joel Spolsky writes about this in his blog post, "Don't Let the Astronaut Architects Scare You" at *https://www .joelonsoftware.com/2001/04/21/dont-let-architecture-astronauts-scare-you/*.

You should know how object-oriented concepts like inheritance work, because they can help you organize your code and make development and debugging easier. Due to Python's flexibility, the language not only offers OOP features, but also doesn't require you to use them when they aren't suited for your program's needs.

Multiple Inheritance

Many programming languages limit classes to at most one parent class. Python supports multiple parent classes by offering a feature called *multiple inheritance*. For example, we can have an Airplane class with a flyInTheAir() method and a Ship class with a floatOnWater() method. We could then create a FlyingBoat class that inherits from both Airplane and Ship by listing both in the class statement, separated by commas. Open a new file editor window and save the following as *flyingboat.py*:

```
class Airplane:
    def flyInTheAir(self):
        print('Flying...')

class Ship:
    def floatOnWater(self):
        print('Floating...')

class FlyingBoat(Airplane, Ship):
    pass
```

The FlyingBoat objects we create will inherit the flyInTheAir() and floatOnWater() methods, as you can see in the interactive shell:

```
>>> from flyingboat import *
>>> seaDuck = FlyingBoat()
>>> seaDuck.flyInTheAir()
Flying...
>>> seaDuck.floatOnWater()
Floating...
```

Multiple inheritance is straightforward as long as the parent classes' method names are distinct and don't overlap. These sorts of classes are called *mixins*. (This is just a general term for a kind of class; Python has no mixin keyword.) But what happens when we inherit from multiple complicated classes that do share method names?

For example, consider the MiniBoard and HintTTTBoard tic-tac-toe board classes from earlier in this chapter. What if we want a class that displays a miniature tic-tac-toe board and also provides hints? With multiple inheritance, we can reuse these existing classes. Add the following to the end of your *tictactoe_oop.py* file but before the if statement that calls the main() function:

```
class HybridBoard(HintBoard, MiniBoard):
    pass
```

This class has nothing in it. It reuses code by inheriting from HintBoard and MiniBoard. Next, change the code in the main() function so it creates a HybridBoard object:

```
gameBoard = HybridBoard() # Create a TTT board object.
```

Both parent classes, MiniBoard and HintBoard, have a method named getBoardStr(), so which one does HybridBoard inherit? When you run this program, the output displays a miniature tic-tac-toe board but also provides hints:

```
--snip--
        X.. 123
        .O. 456
        X.. 789
X can win in one more move.
```

Python seems to have magically merged the MiniBoard class's getBoardStr() method and the HintBoard class's getBoardStr() method to do both! But this is because I've written them to work with each other. In fact, if you switch the order of the classes in the HybridBoard class's class statement so it looks like this:

```
class HybridBoard(MiniBoard, HintBoard):
```

you lose the hints altogether:

```
--snip--
        X.. 123
        .O. 456
        X.. 789
```

To understand why this happens, you need to understand Python's *method resolution order* (*MRO*) and how the super() function actually works.

Method Resolution Order

Our tic-tac-toe program now has four classes to represent boards, three with defined getBoardStr() methods and one with an inherited getBoardStr() method, as shown in Figure 16-2.

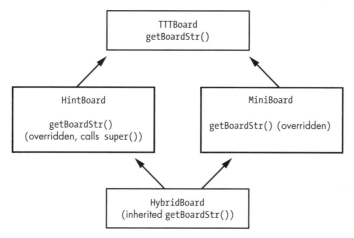

Figure 16-2: The four classes in our tic-tac-toe board program

When we call getBoardStr() on a HybridBoard object, Python knows that the HybridBoard class doesn't have a method with this name, so it checks its parent class. But the class has two parent classes, both of which have a getBoardStr() method. Which one gets called?

You can find out by checking the HybridBoard class's MRO, which is the ordered list of classes that Python checks when inheriting methods or when a method calls the super() function. You can see the HybridBoard class's MRO by calling its mro() method in the interactive shell:

```
>>> from tictactoe_oop import *
>>> HybridBoard.mro()
[<class 'tictactoe_oop.HybridBoard'>, <class 'tictactoe_oop.HintBoard'>,
<class 'tictactoe_oop.MiniBoard'>, <class 'tictactoe_oop.TTTBoard'>, <class
'object'>]
```

From this return value, you can see that when a method is called on HybridBoard, Python first checks for it in the HybridBoard class. If it's not there, Python checks the HintBoard class, then the MiniBoard class, and finally the TTTBoard class. At the end of every MRO list is the built-in object class, which is the parent class of all classes in Python.

For single inheritance, determining the MRO is easy: just make a chain of parent classes. For multiple inheritance, it's trickier. Python's MRO follows the C3 algorithm, whose details are beyond the scope of this book. But you can determine the MRO by remembering two rules:

- Python checks child classes before parent classes.
- Python checks inherited classes listed left to right in the class statement.

If we call getBoardStr() on a HybridBoard object, Python checks the HybridBoard class first. Then, because the class's parents from left to right are HintBoard and MiniBoard, Python checks HintBoard. This parent class has a getBoardStr() method, so HybridBoard inherits and calls it.

But it doesn't end there: next, this method calls super().getBoardStr(). *Super* is a somewhat misleading name for Python's super() function, because it doesn't return the parent class but rather the next class in the MRO. This means that when we call getBoardStr() on a HybridBoard object, the next class in its MRO, after HintBoard, is MiniBoard, not the parent class TTTBoard. So the call to super().getBoardStr() calls the MiniBoard class's getBoardStr() method, which returns the miniature tic-tac-toe board string. The remaining code in the HintBoard class's getBoardStr() after this super() call appends the hint text to this string.

If we change the HybridBoard class's class statement so it lists MiniBoard first and HintBoard second, its MRO will put MiniBoard before HintBoard. This means HybridBoard inherits getBoardStr() from MiniBoard, which doesn't have a call to super(). This ordering is what caused the bug that made the miniature tic-tac-toe board display without hints: without a super() call, the MiniBoard class's getBoardStr() method never calls the HintBoard class's getBoardStr() method.

Multiple inheritance allows you to create a lot of functionality in a small amount of code but easily leads to overengineered, hard-to-understand code. Favor single inheritance, mixin classes, or no inheritance. These techniques are often more than capable of carrying out your program's tasks.

Summary

Inheritance is a technique for code reuse. It lets you create child classes that inherit the methods of their parent classes. You can override the methods to provide new code for them but also use the super() function to call the original methods in the parent class. A child class has an "is a" relationship with its parent class, because an object of the child class is a kind of object of the parent class.

In Python, using classes and inheritance is optional. Some programmers see the complexity that heavy use of inheritance creates as not worth its benefits. It's often more flexible to use composition instead of inheritance, because it implements a "has a" relationship with an object of one class and objects of other classes rather than inheriting methods directly from those other classes. This means that objects of one class can have an object of another class. For example, a Customer object could have a birthdate method that is assigned a Date object rather than the Customer class subclassing Date.

Just as type() can return the type of the object passed to it, the isinstance() and issubclass() functions return type and inheritance information about the object passed to them.

Classes can have object methods and attributes, but they can also have class methods, class attributes, and static methods. Although these are rarely used, they can enable other object-oriented techniques that global variables and functions can't provide.

Python lets classes inherit from multiple parents, although this can lead to code that is difficult to understand. The super() function and a class's methods figure out how to inherit methods based on the MRO. You can view a class's MRO in the interactive shell by calling the mro() method on the class.

This chapter and the previous one covered general OOP concepts. In the next chapter, we'll explore Python-specific OOP techniques.

17

PYTHONIC OOP: PROPERTIES AND DUNDER METHODS

Many languages have OOP features, but Python has some unique OOP features, including properties and dunder methods. Learning how to use these Pythonic techniques can help you write concise and readable code.

Properties allow you to run some specific code each time an object's attribute is read, modified, or deleted to ensure the object isn't put into an invalid state. In other languages, this code is often called *getters* or *setters*. Dunder methods allow you to use your objects with Python's built-in operators, such as the + operator. This is how you can combine two datetime.timedelta objects, such as datetime.timedelta(days=2) and datetime.timedelta(days=3), to create a new datetime.timedelta(days=5) object.

In addition to using other examples, we'll continue to expand the WizCoin class we started in Chapter 15 by adding properties and overloading operators with dunder methods. These features will make WizCoin objects more expressive and easier to use in any application that imports the wizcoin module.

Properties

The BankAccount class that we used in Chapter 15 marked its _balance attribute as private by placing an underscore at the start of its name. But remember that designating an attribute as private is only a convention: all attributes in Python are technically public, meaning they're accessible to code outside the class. There's nothing to prevent code from intentionally or maliciously changing the _balance attribute to an invalid value.

But you can prevent *accidental* invalid changes to these private attributes with properties. In Python, *properties* are attributes that have specially assigned *getter*, *setter*, and *deleter* methods that can regulate how the attribute is read, changed, and deleted. For example, if the attribute is only supposed to have integer values, setting it to the string '42' will likely cause bugs. A property would call the setter method to run code that fixes, or at least provides early detection of, setting an invalid value. If you've thought, "I wish I could run some code each time this attribute was accessed, modified with an assignment statement, or deleted with a del statement," then you want to use properties.

Turning an Attribute into a Property

First, let's create a simple class that has a regular attribute instead of a property. Open a new file editor window and enter the following code, saving it as *regularAttributeExample.py*:

```
class ClassWithRegularAttributes:
    def __init__(self, someParameter):
        self.someAttribute = someParameter

obj = ClassWithRegularAttributes('some initial value')
print(obj.someAttribute)  # Prints 'some initial value'
obj.someAttribute = 'changed value'
print(obj.someAttribute)  # Prints 'changed value'
del obj.someAttribute  # Deletes the someAttribute attribute.
```

This ClassWithRegularAttributes class has a regular attribute named someAttribute. The __init__() method sets someAttribute to 'some initial value', but we then directly change the attribute's value to 'changed value'. When you run this program, the output looks like this:

```
some initial value
changed value
```

This output indicates that code can easily change someAttribute to any value. The downside of using regular attributes is that your code can set the someAttribute attribute to invalid values. This flexibility is simple and convenient, but it also means someAttribute could be set to some invalid value that causes bugs.

Let's rewrite this class using properties by following these steps to do this for an attribute named someAttribute:

1. Rename the attribute with an underscore prefix: _someAttribute.
2. Create a method named someAttribute with the @property decorator. This getter method has the self parameter that all methods have.
3. Create another method named someAttribute with the @someAttribute .setter decorator. This setter method has parameters named self and value.
4. Create another method named someAttribute with the @someAttribute .deleter decorator. This deleter method has the self parameter that all methods have.

Open a new file editor window and enter the following code, saving it as *propertiesExample.py*:

```
class ClassWithProperties:
    def __init__(self):
        self.someAttribute = 'some initial value'

    @property
    def someAttribute(self): # This is the "getter" method.
        return self._someAttribute

    @someAttribute.setter
    def someAttribute(self, value): # This is the "setter" method.
        self._someAttribute = value

    @someAttribute.deleter
    def someAttribute(self): # This is the "deleter" method.
        del self._someAttribute

obj = ClassWithProperties()
print(obj.someAttribute) # Prints 'some initial value'
obj.someAttribute = 'changed value'
print(obj.someAttribute) # Prints 'changed value'
del obj.someAttribute # Deletes the _someAttribute attribute.
```

This program's output is the same as the *regularAttributeExample.py* code, because they effectively do the same task: they print an object's initial attribute and then update that attribute and print it again.

But notice that the code outside the class never directly accesses the _someAttribute attribute (it's private, after all). Instead, the outside code accesses the someAttribute property. What this property actually consists of is a bit abstract: the getter, setter, and deleter methods combined make up the property. When we rename an attribute named someAttribute to _someAttribute while creating getter, setter, and deleter methods for it, we call this the someAttribute property.

In this context, the _someAttribute attribute is called a *backing field* or *backing variable* and is the attribute on which the property is based. Most, but not all, properties use a backing variable. We'll create a property without a backing variable in "Read-Only Properties" later in this chapter.

You never call the getter, setter, and deleter methods in your code because Python does it for you under the following circumstances:

- When Python runs code that accesses a property, such as print(obj.someAttribute), behind the scenes, it calls the getter method and uses the returned value.
- When Python runs an assignment statement with a property, such as obj.someAttribute = 'changed value', behind the scenes, it calls the setter method, passing the 'changed value' string for the value parameter.
- When Python runs a del statement with a property, such as del obj.someAttribute, behind the scenes, it calls the deleter method.

The code in the property's getter, setter, and deleter methods acts on the backing variable directly. You don't want the getter, setter, or deleter methods to act on the property, because this could cause errors. In one possible example, the getter method would access the property, causing the getter method to call itself, which makes it access the property again, causing it to call itself again, and so on until the program crashes. Open a new file editor window and enter the following code, saving it as *badPropertyExample.py*:

```python
class ClassWithBadProperty:
    def __init__(self):
        self.someAttribute = 'some initial value'

    @property
    def someAttribute(self):  # This is the "getter" method.
        # We forgot the _ underscore in `self._someAttribute here`, causing
        # us to use the property and call the getter method again:
        return self.someAttribute  # This calls the getter again!

    @someAttribute.setter
    def someAttribute(self, value):  # This is the "setter" method.
        self._someAttribute = value

obj = ClassWithBadProperty()
print(obj.someAttribute)  # Error because the getter calls the getter.
```

When you run this code, the getter continually calls itself until Python raises a RecursionError exception:

```
Traceback (most recent call last):
  File "badPropertyExample.py", line 16, in <module>
    print(obj.someAttribute)  # Error because the getter calls the getter.
  File "badPropertyExample.py", line 9, in someAttribute
    return self.someAttribute  # This calls the getter again!
```

```
      File "badPropertyExample.py", line 9, in someAttribute
        return self.someAttribute  # This calls the getter again!
      File "badPropertyExample.py", line 9, in someAttribute
        return self.someAttribute  # This calls the getter again!
      [Previous line repeated 996 more times]
    RecursionError: maximum recursion depth exceeded
```

To prevent this recursion, the code inside your getter, setter, and deleter methods should always act on the backing variable (which should have an underscore prefix in its name), never the property. Code outside these methods should use the property, although as with the private access underscore prefix convention, nothing prevents you from writing code on the backing variable anyway.

Using Setters to Validate Data

The most common need for using properties is to *validate* data or to make sure it's in the format you want it to be in. You might not want code outside the class to be able to set an attribute to just any value; this could lead to bugs. You can use properties to add checks that ensure only valid values are assigned to an attribute. These checks let you catch bugs earlier in code development, because they raise an exception as soon as an invalid value is set.

Let's update the *wizcoin.py* file from Chapter 15 to turn the galleons, sickles, and knuts attributes into properties. We'll change the setter for these properties so only positive integers are valid. Our WizCoin objects represent an amount of coins, and you can't have half a coin or an amount of coins less than zero. If code outside the class tries to set the galleons, sickles, or knuts properties to an invalid value, we'll raise a WizCoinException exception.

Open the *wizcoin.py* file that you saved in Chapter 15 and modify it to look like the following:

❶ ```
  class WizCoinException(Exception):
```
❷ ```
      """The wizcoin module raises this when the module is misused."""
      pass

  class WizCoin:
      def __init__(self, galleons, sickles, knuts):
          """Create a new WizCoin object with galleons, sickles, and knuts."""
```
❸ ```
          self.galleons = galleons
          self.sickles = sickles
          self.knuts   = knuts
          # NOTE: __init__() methods NEVER have a return statement.

  --snip--

      @property
```
❹ ```
      def galleons(self):
          """Returns the number of galleon coins in this object."""
          return self._galleons
```

```
        @galleons.setter
❺    def galleons(self, value):
❻        if not isinstance(value, int):
❼            raise WizCoinException('galleons attr must be set to an int, not a
' + value.__class__.__qualname__)
❽        if value < 0:
             raise WizCoinException('galleons attr must be a positive int, not
' + value.__class__.__qualname__)
         self._galleons = value

--snip--
```

The new changes add a WizCoinException class ❶ that inherits from Python's built-in Exception class. The class's docstring describes how the wizcoin module ❷ uses it. This is a best practice for Python modules: the WizCoin class's objects can raise this when they're misused. That way, if a WizCoin object raises other exception classes, like ValueError or TypeError, this will mostly likely signify that it's a bug in the WizCoin class.

In the __init__() method, we set the self.galleons, self.sickles, and self.knuts properties ❸ to the corresponding parameters.

At the bottom of the file, after the total() and weight() methods, we add a getter ❹ and setter method ❺ for the self._galleons attribute. The getter simply returns the value in self._galleons. The setter checks whether the value being assigned to the galleons property is an integer ❻ and positive ❽. If either check fails, WizCoinException is raised with an error message. This check prevents _galleons from ever being set with an invalid value as long as code always uses the galleons property.

All Python objects automatically have a __class__ attribute, which refers to the object's class object. In other words, value.__class__ is the same class object that type(value) returns. This class object has an attribute named __qualname__ that is a string of the class's name. (Specifically, it's the *qualified* name of the class, which includes the names of any classes the class object is nested in. Nested classes are of limited use and beyond the scope of this book.) For example, if value stored the date object returned by datetime.date(2021, 1, 1), then value.__class__.__qualname__ would be the string 'date'. The exception messages use value.__class__.__qualname__ ❼ to get a string of the value object's name. The class name makes the error message more useful to the programmer reading it, because it identifies not only that the value argument was not the right type, but what type it was and what type it should be.

You'll need to copy the code for the getter and setter for _galleons to use for the _sickles and _knuts attributes as well. Their code is identical except they use the _sickles and _knuts attributes, instead of _galleons, as backing variables.

Read-Only Properties

Your objects might need some read-only properties that can't be set with the assignment operator =. You can make a property read-only by omitting the setter and deleter methods.

For example, the `total()` method in the `WizCoin` class returns the value of the object in knuts. We could change this from a regular method to a read-only property, because there is no reasonable way to set the total of a `WizCoin` object. After all, if you set `total` to the integer 1000, does this mean 1,000 knuts? Or does it mean 1 galleon and 493 knuts? Or does it mean some other combination? For this reason, we'll make `total` a read-only property by adding the code in bold to the *wizcoin.py* file:

```python
@property
def total(self):
    """Total value (in knuts) of all the coins in this WizCoin object."""
    return (self.galleons * 17 * 29) + (self.sickles * 29) + (self.knuts)

# Note that there is no setter or deleter method for `total`.
```

After you add the `@property` function decorator in front of `total()`, Python will call the `total()` method whenever `total` is accessed. Because there is no setter or deleter method, Python raises `AttributeError` if any code attempts to modify or delete `total` by using it in an assignment or `del` statement, respectively. Notice that the value of the `total` property depends on the value in the `galleons`, `sickles`, and `knuts` properties: this property isn't based on a backing variable named `_total`. Enter the following into the interactive shell:

```python
>>> import wizcoin
>>> purse = wizcoin.WizCoin(2, 5, 10)
>>> purse.total
1141
>>> purse.total = 1000
Traceback (most recent call last):
  File "<stdin>", line 1, in <module>
AttributeError: can't set attribute
```

You might not like that your program immediately crashes when you attempt to change a read-only property, but this behavior is preferable to allowing a change to a read-only property. Your program being able to modify a read-only property would certainly cause a bug at some point while the program runs. If this bug happens much later after you modify the read-only property, it would be hard to track down the original cause. Crashing immediately allows you to notice the problem sooner.

Don't confuse read-only properties with constant variables. Constant variables are written in all uppercase and rely on the programmer to not modify them. Their value is supposed to remain constant and unchanging for the duration of a program's run. A read-only property is, as with any attribute, associated with an object. A read-only property cannot be directly set or deleted. But it might evaluate to a changing value. Our `WizCoin` class's `total` property changes as its `galleons`, `sickles`, and `knuts` properties change.

When to Use Properties

As you saw in the previous sections, properties provide more control over how we can use a class's attributes, and they're a Pythonic way to write code. Methods with names like getSomeAttribute() or setSomeAttribute() signal that you should probably use properties instead.

This isn't to say that *every* instance of a method beginning with *get* or *set* should immediately be replaced with a property. There are situations in which you should use a method, even if its name begins with *get* or *set*. Here are some examples:

- For slow operations that take more than a second or two—for example, downloading or uploading a file

- For operations that have side effects, such as changes to other attributes or objects

- For operations that require additional arguments to be passed to the get or set operation—for example, in a method call like emailObj .getFileAttachment(filename)

Programmers often think of methods as verbs (in the sense that methods perform some action), and they think of attributes and properties as nouns (in the sense that they represent some item or object). If your code seems to be performing more of an action of getting or setting rather than getting or setting an item, it might be best to use a getter or setter method. Ultimately, this decision depends on what sounds right to you as the programmer.

The great advantage of using Python's properties is that you don't have to use them when you first create your class. You can use regular attributes, and if you need properties later, you can convert the attributes to properties without breaking any code outside the class. When we make a property with the attribute's name, we can rename the attribute using a prefix underscore and our program will still work as it did before.

Python's Dunder Methods

Python has several special method names that begin and end with double underscores, abbreviated as *dunder*. These methods are called *dunder methods*, *special methods*, or *magic methods*. You're already familiar with the __init__() dunder method name, but Python has several more. We often use them for *operator overloading*—that is, adding custom behaviors that allow us to use objects of our classes with Python operators, such as + or >=. Other dunder methods let objects of our classes work with Python's built-in functions, such as len() or repr().

As with __init__() or the getter, setter, and deleter methods for properties, you almost never call dunder methods directly. Python calls them behind the scenes when you use the objects with operators or built-in functions. For example, if you create a method named __len__() or __repr__() for your class, they'll be called behind the scenes when an object of that class is passed to the len() or repr() function, respectively. These methods are documented online in the official Python documentation at *https://docs.python .org/3/reference/datamodel.html*.

As we explore the many different types of dunder methods, we'll expand our WizCoin class to take advantage of them.

String Representation Dunder Methods

You can use the __repr_() and __str_() dunder methods to create string representations of objects that Python typically doesn't know how to handle. Usually, Python creates string representations of objects in two ways. The *repr* (pronounced "repper") string is a string of Python code that, when run, creates a copy of the object. The *str* (pronounced "stir") string is a human-readable string that provides clear, useful information about the object. The repr and str strings are returned by the repr() and str() built-in functions, respectively. For example, enter the following into the interactive shell to see a datetime.date object's repr and str strings:

```
>>> import datetime
❶ >>> newyears = datetime.date(2021, 1, 1)
>>> repr(newyears)
❷ 'datetime.date(2021, 1, 1)'
>>> str(newyears)
❸ '2021-01-01'
❹ >>> newyears
datetime.date(2021, 1, 1)
```

In this example, the 'datetime.date(2021, 1, 1)' repr string of the date time.date object ❷ is literally a string of Python code that creates a copy of that object ❶. This copy provides a precise representation of the object. On the other hand, the '2021-01-01' str string of the datetime.date object ❸ is a string representing the object's value in a way that's easy for humans to read. If we simply enter the object into the interactive shell ❹, it displays the repr string. An object's str string is often displayed to users, whereas an object's repr string is used in technical contexts, such as error messages and logfiles.

Python knows how to display objects of its built-in types, such as integers and strings. But it can't know how to display objects of the classes we create. If repr() doesn't know how to create a repr or str string for an object, by convention the string will be enclosed in angle brackets and contain the object's memory address and class name: '<wizcoin.WizCoin object at 0x00000212B4148EE0>'. To create this kind of string for a WizCoin object, enter the following into the interactive shell:

```
>>> import wizcoin
>>> purse = wizcoin.WizCoin(2, 5, 10)
>>> str(purse)
'<wizcoin.WizCoin object at 0x00000212B4148EE0>'
>>> repr(purse)
'<wizcoin.WizCoin object at 0x00000212B4148EE0>'
>>> purse
<wizcoin.WizCoin object at 0x00000212B4148EE0>
```

These strings aren't very readable or useful, so we can tell Python what strings to use by implementing the __repr__() and __str__() dunder methods. The __repr__() method specifies what string Python should return when the object is passed to the repr() built-in function, and the __str__() method specifies what string Python should return when the object is passed to the str() built-in function. Add the following to the end of the *wizcoin.py* file:

```
--snip--
    def __repr__(self):
        """Returns a string of an expression that re-creates this object."""
        return f'{self.__class__.__qualname__}({self.galleons}, {self.sickles}, {self.knuts})'

    def __str__(self):
        """Returns a human-readable string representation of this object."""
        return f'{self.galleons}g, {self.sickles}s, {self.knuts}k'
```

When we pass purse to repr() and str(), Python calls the __repr__() and __str__() dunder methods. We don't call the dunder methods in our code.

Note that f-strings that include the object in braces will implicitly call str() to get an object's str string. For example, enter the following into the interactive shell:

```
>>> import wizcoin
>>> purse = wizcoin.WizCoin(2, 5, 10)
>>> repr(purse)  # Calls WizCoin's __repr__() behind the scenes.
'WizCoin(2, 5, 10)'
>>> str(purse)  # Calls WizCoin's __str__() behind the scenes.
'2g, 5s, 10k'
>>> print(f'My purse contains {purse}.')  # Calls WizCoin's __str__().
My purse contains 2g, 5s, 10k.
```

When we pass the WizCoin object in purse to the repr() and str() functions, behind the scenes Python calls the WizCoin class's __repr__() and __str__() methods. We programmed these methods to return more readable and useful strings. If you entered the text of the 'WizCoin(2, 5, 10)' repr string into the interactive shell, it would create a WizCoin object that has the same attributes as the object in purse. The str string is a more human-readable representation of the object's value: '2g, 5s, 10k'. If you use a WizCoin object in an f-string, Python uses the object's str string.

If WizCoin objects were so complex that it would be impossible to create a copy of them with a single constructor function call, we would enclose the repr string in angle brackets to denote that it's not meant to be Python code. This is what the generic representation strings, such as '<wizcoin. WizCoin object at 0x00000212B4148EE0>', do. Typing this string into the interactive shell would raise a SyntaxError, so it couldn't possibly be confused for Python code that creates a copy of the object.

Inside the __repr__() method, we use self.__class__.__qualname__ instead of hardcoding the string 'WizCoin'; so if we subclass WizCoin, the inherited __repr__() method will use the subclass's name instead of 'WizCoin'. In addition, if we rename the WizCoin class, the __repr__() method will automatically use the updated name.

But the `WizCoin` object's str string shows us the attribute values in a neat, concise form. I highly recommended you implement __repr__() and __str__() in all your classes.

SENSITIVE INFORMATION IN REPR STRINGS

As mentioned earlier, we usually display the str string to users, and we use the repr string in technical contexts, such as logfiles. But the repr string can cause security issues if the object you're creating contains sensitive information, such as passwords, medical details, or personally identifiable information. If this is the case, make sure the __repr__() method doesn't include this information in the string it returns. When software crashes, it's frequently set up to include the contents of variables in a logfile to aid in debugging. Often, these logfiles aren't treated as sensitive information. In several security incidents, publicly shared logfiles have inadvertently included passwords, credit card numbers, home addresses, and other sensitive information. Keep this in mind when you're writing __repr__() methods for your class.

Numeric Dunder Methods

The *numeric dunder methods*, also called the *math dunder methods*, overload Python's mathematical operators, such as +, -, *, /, and so on. Currently, we can't perform an operation like adding two `WizCoin` objects together with the + operator. If we try to do so, Python will raise a `TypeError` exception, because it doesn't know how to add `WizCoin` objects. To see this error, enter the following into the interactive shell:

```
>>> import wizcoin
>>> purse = wizcoin.WizCoin(2, 5, 10)
>>> tipJar = wizcoin.WizCoin(0, 0, 37)
>>> purse + tipJar
Traceback (most recent call last):
  File "<stdin>", line 1, in <module>
TypeError: unsupported operand type(s) for +: 'WizCoin' and 'WizCoin'
```

Instead of writing an `addWizCoin()` method for the `WizCoin` class, you can use the __add__() dunder method so `WizCoin` objects work with the + operator. Add the following to the end of the *wizcoin.py* file:

```
--snip--
❶   def __add__(self, other):
        """Adds the coin amounts in two WizCoin objects together."""
❷       if not isinstance(other, WizCoin):
         return NotImplemented

❸       return WizCoin(other.galleons + self.galleons, other.sickles +
        self.sickles, other.knuts + self.knuts)
```

When a WizCoin object is on the left side of the + operator, Python calls the __add__() method ❶ and passes in the value on the right side of the + operator for the other parameter. (The parameter can be named anything, but other is the convention.)

Keep in mind that you can pass any type of object to the __add__() method, so the method must include type checks ❷. For example, it doesn't make sense to add an integer or a float to a WizCoin object, because we don't know whether it should be added to the galleons, sickles, or knuts amount.

The __add__() method creates a new WizCoin object with amounts equal to the sum of the galleons, sickles, and knuts attributes of self and other ❸. Because these three attributes contain integers, we can use the + operator on them. Now that we've overloaded the + operator for the WizCoin class, we can use the + operator on WizCoin objects.

Overloading the + operator like this allows us to write more readable code. For example, enter the following into the interactive shell:

```
>>> import wizcoin
>>> purse = wizcoin.WizCoin(2, 5, 10)  # Create a WizCoin object.
>>> tipJar = wizcoin.WizCoin(0, 0, 37)  # Create another WizCoin object.
>>> purse + tipJar  # Creates a new WizCoin object with the sum amount.
WizCoin(2, 5, 47)
```

If the wrong type of object is passed for other, the dunder method shouldn't raise an exception but rather return the built-in value NotImplemented. For example, in the following code, other is an integer:

```
>>> import wizcoin
>>> purse = wizcoin.WizCoin(2, 5, 10)
>>> purse + 42  # WizCoin objects and integers can't be added together.
Traceback (most recent call last):
  File "<stdin>", line 1, in <module>
TypeError: unsupported operand type(s) for +: 'WizCoin' and 'int'
```

Returning NotImplemented signals Python to try calling other methods to perform this operation. (See "Reflected Numeric Dunder Methods" later in this chapter for more details.) Behind the scenes, Python calls the __add__() method with 42 for the other parameter, which also returns NotImplemented, causing Python to raise a TypeError.

Although we shouldn't be able to add integers to or subtract them from WizCoin objects, it would make sense to allow code to multiply WizCoin objects by positive integer amounts by defining a __mul__() dunder method. Add the following to the end of *wizcoin.py*:

```
--snip--
    def __mul__(self, other):
        """Multiplies the coin amounts by a non-negative integer."""
        if not isinstance(other, int):
            return NotImplemented
        if other < 0:
            # Multiplying by a negative int results in negative
            # amounts of coins, which is invalid.
```

```
        raise WizCoinException('cannot multiply with negative integers')

    return WizCoin(self.galleons * other, self.sickles * other, self.knuts * other)
```

This __mul__() method lets you multiply WizCoin objects by positive integers. If other is an integer, it's the data type the __mul__() method is expecting and we shouldn't return NotImplemented. But if this integer is negative, multiplying the WizCoin object by it would result in negative amounts of coins in our WizCoin object. Because this goes against our design for this class, we raise a WizCoinException with a descriptive error message.

NOTE   *You shouldn't change the self object in a numeric dunder method. Rather, the method should always create and return a new object. The + and other numeric operators are always expected to evaluate to a new object rather than modifying an object's value in-place.*

Enter the following into the interactive shell to see the __mul__() dunder method in action:

```
>>> import wizcoin
>>> purse = wizcoin.WizCoin(2, 5, 10)  # Create a WizCoin object.
>>> purse * 10  # Multiply the WizCoin object by an integer.
WizCoin(20, 50, 100)
>>> purse * -2  # Multiplying by a negative integer causes an error.
Traceback (most recent call last):
  File "<stdin>", line 1, in <module>
  File "C:\Users\Al\Desktop\wizcoin.py", line 86, in __mul__
    raise WizCoinException('cannot multiply with negative integers')
wizcoin.WizCoinException: cannot multiply with negative integers
```

Table 17-1 shows the full list of numeric dunder methods. You don't always need to implement all of them for your class. It's up to you to decide which methods are relevant.

Table 17-1: Numeric Dunder Methods

Dunder method	Operation	Operator or built-in function
__add__()	Addition	+
__sub__()	Subtraction	-
__mul__()	Multiplication	*
__matmul__()	Matrix multiplication (new in Python 3.5)	@
__truediv__()	Division	/
__floordiv__()	Integer division	//
__mod__()	Modulus	%
__divmod__()	Division and modulus	divmod()

(continued)

Table 17-1: Numeric Dunder Methods *(continued)*

Dunder method	Operation	Operator or built-in function
__pow__()	Exponentiation	**, pow()
__lshift__()	Left shift	>>
__rshift__()	Right shift	<<
__and__()	Bitwise and	&
__or__()	Bitwise or	|
__xor__()	Bitwise exclusive or	^
__neg__()	Negation	Unary -, as in -42
__pos__()	Identity	Unary +, as in +42
__abs__()	Absolute value	abs()
__invert__()	Bitwise inversion	~
__complex__()	Complex number form	complex()
__int__()	Integer number form	int()
__float__()	Floating-point number form	float()
__bool__()	Boolean form	bool()
__round__()	Rounding	round()
__trunc__()	Truncation	math.trunc()
__floor__()	Rounding down	math.floor()
__ceil__()	Rounding up	math.ceil()

Some of these methods are relevant to our `WizCoin` class. Try writing your own implementation of the __sub__(), __pow__(), __int__(), __float__(), and __bool__() methods. You can see an example of an implementation at *https://autbor.com/wizcoinfull*. The full documentation for the numeric dunder methods is in the Python documentation at *https://docs.python.org/3/reference/datamodel.html#emulating-numeric-types*.

The numeric dunder methods allow objects of your classes to use Python's built-in math operators. If you're writing methods with names like `multiplyBy()`, `convertToInt()`, or something similar that describes a task typically done by an existing operator or built-in function, use the numeric dunder methods (as well as the reflected and in-place dunder methods described in the next two sections).

Reflected Numeric Dunder Methods

Python calls the numeric dunder methods when the object is on the left side of a math operator. But it calls the *reflected* numeric dunder methods (also called the *reverse* or *right-hand* dunder methods) when the object is on the right side of a math operator.

Reflected numeric dunder methods are useful because programmers using your class won't always write the object on the left side of the operator, which could lead to unexpected behavior. For example, let's consider what happens when purse contains a WizCoin object, and Python evaluates the expression 2 * purse, where purse is on the right side of the operator:

1. Because 2 is an integer, the int class's __mul__() method is called with purse passed for the other parameter.
2. The int class's __mul__() method doesn't know how to handle WizCoin objects, so it returns NotImplemented.
3. Python doesn't raise a TypeError just yet. Because purse contains a WizCoin object, the WizCoin class's __rmul__() method is called with 2 passed for the other parameter.
4. If __rmul__() returns NotImplemented, Python raises a TypeError.

Otherwise, the returned object from __rmul__() is what the 2 * purse expression evaluates to.

But the expression purse * 2, where purse is on the left side of the operator, works differently:

1. Because purse contains a WizCoin object, the WizCoin class's __mul__() method is called with 2 passed for the other parameter.
2. The __mul__() method creates a new WizCoin object and returns it.
3. This returned object is what the purse * 2 expression evaluates to.

Numeric dunder methods and reflected numeric dunder methods have identical code if they are *commutative*. Commutative operations, like addition, have the same result backward and forward: 3 + 2 is the same as 2 + 3. But other operations aren't commutative: 3 − 2 is not the same as 2 − 3. Any commutative operation can just call the original numeric dunder method whenever the reflected numeric dunder method is called. For example, add the following to the end of the *wizcoin.py* file to define a reflected numeric dunder method for the multiplication operation:

```
--snip--
    def __rmul__(self, other):
        """Multiplies the coin amounts by a non-negative integer."""
        return self.__mul__(other)
```

Multiplying an integer and a WizCoin object is commutative: 2 * purse is the same as purse * 2. Instead of copying and pasting the code from __mul__(), we just call self.__mul__() and pass it the other parameter.

After updating *wizcoin.py*, practice using the reflected multiplication dunder method by entering the following into the interactive shell:

```
>>> import wizcoin
>>> purse = wizcoin.WizCoin(2, 5, 10)
>>> purse * 10  # Calls __mul__() with 10 for the `other` parameter.
WizCoin(20, 50, 100)
>>> 10 * purse  # Calls __rmul__() with 10 for the `other` parameter.
WizCoin(20, 50, 100)
```

Keep in mind that in the expression 10 * purse, Python first calls the int class's _mul_() method to see whether integers can be multiplied with WizCoin objects. Of course, Python's built-in int class doesn't know anything about the classes we create, so it returns NotImplemented. This signals to Python to next call WizCoin class's _rmul_(), and if it exists, to handle this operation. If the calls to the int class's _mul_() and WizCoin class's _rmul_() both return NotImplemented, Python raises a TypeError exception.

Only WizCoin objects can be added to each other. This guarantees that the first WizCoin object's _add_() method will handle the operation, so we don't need to implement _radd_(). For example, in the expression purse + tipJar, the _add_() method for the purse object is called with tipJar passed for the other parameter. Because this call won't return NotImplemented, Python doesn't try to call the tipJar object's _radd_() method with purse as the other parameter.

Table 17-2 contains a full listing of the available reflected dunder methods.

Table 17-2: Reflected Numeric Dunder Methods

Dunder method	Operation	Operator or built-in function
radd()	Addition	+
rsub()	Subtraction	-
rmul()	Multiplication	*
rmatmul()	Matrix multiplication (new in Python 3.5)	@
rtruediv()	Division	/
rfloordiv()	Integer division	//
rmod()	Modulus	%
rdivmod()	Division and modulus	divmod()
rpow()	Exponentiation	**, pow()
rlshift()	Left shift	>>
rrshift()	Right shift	<<
rand()	Bitwise and	&
ror()	Bitwise or		
rxor()	Bitwise exclusive or	^

The full documentation for the reflected dunder methods is in the Python documentation at *https://docs.python.org/3/reference/datamodel.html #emulating-numeric-types*.

In-Place Augmented Assignment Dunder Methods

The numeric and reflected dunder methods always create new objects rather than modifying the object in-place. The *in-place dunder methods*, called by the augmented assignment operators, such as += and *=, modify the object in-place rather than creating new objects. (There is an exception to this, which I'll explain at the end of this section.) These dunder method names begin with an *i*, such as __iadd__() and __imul__() for the += and *= operators, respectively.

For example, when Python runs the code purse *= 2, the expected behavior isn't that the WizCoin class's __imul__() method creates and returns a new WizCoin object with twice as many coins, and then assigns it the purse variable. Instead, the __imul__() method modifies the existing WizCoin object in purse so it has twice as many coins. This is a subtle but important difference if you want your classes to overload the augmented assignment operators.

Our WizCoin objects already overload the + and * operators, so let's define the __iadd__() and __imul__() dunder methods so they overload the += and *= operators as well. In the expressions purse += tipJar and purse *= 2, we call the __iadd__() and __imul__() methods, respectively, with tipJar and 2 passed for the other parameter, respectively. Add the following to the end of the *wizcoin.py* file:

```
--snip--
    def __iadd__(self, other):
        """Add the amounts in another WizCoin object to this object."""
        if not isinstance(other, WizCoin):
            return NotImplemented

        # We modify the `self` object in-place:
        self.galleons += other.galleons
        self.sickles += other.sickles
        self.knuts += other.knuts
        return self  # In-place dunder methods almost always return self.

    def __imul__(self, other):
        """Multiply the amount of galleons, sickles, and knuts in this object
        by a non-negative integer amount."""
        if not isinstance(other, int):
            return NotImplemented
        if other < 0:
            raise WizCoinException('cannot multiply with negative integers')

        # The WizCoin class creates mutable objects, so do NOT create a
        # new object like this commented-out code:
        #return WizCoin(self.galleons * other, self.sickles * other, self.knuts * other)

        # We modify the `self` object in-place:
        self.galleons *= other
        self.sickles *= other
        self.knuts *= other
        return self  # In-place dunder methods almost always return self.
```

The WizCoin objects can use the += operator with other WizCoin objects and the *= operator with positive integers. Notice that after ensuring that the other parameter is valid, the in-place methods modify the self object in-place rather than creating a new WizCoin object. Enter the following into the interactive shell to see how the augmented assignment operators modify the WizCoin objects in-place:

```
>>> import wizcoin
>>> purse = wizcoin.WizCoin(2, 5, 10)
>>> tipJar = wizcoin.WizCoin(0, 0, 37)
❶ >>> purse + tipJar
❷ WizCoin(2, 5, 46)
>>> purse
WizCoin(2, 5, 10)
❸ >>> purse += tipJar
>>> purse
WizCoin(2, 5, 47)
❹ >>> purse *= 10
>>> purse
WizCoin(20, 50, 470)
```

The + operator ❶ calls the __add__() or __radd__() dunder methods to create and return new objects ❷. The original objects operated on by the + operator remain unmodified. The in-place dunder methods ❸ ❹ should modify the object in-place as long as the object is mutable (that is, it's an object whose value can change). The exception is for immutable objects: because an immutable object can't be modified, it's impossible to modify it in-place. In that case, the in-place dunder methods should create and return a new object, just like the numeric and reflected numeric dunder methods.

We didn't make the galleons, sickles, and knuts attributes read-only, which means they can change. So WizCoin objects are mutable. Most of the classes you write will create mutable objects, so you should design your in-place dunder methods to modify the object in-place.

If you don't implement an in-place dunder method, Python will instead call the numeric dunder method. For example, if the WizCoin class had no __imul__() method, the expression purse *= 10 will call __mul__() instead and assign its return value to purse. Because WizCoin objects are mutable, this is unexpected behavior that could lead to subtle bugs.

Comparison Dunder Methods

Python's sort() method and sorted() function contain an efficient sorting algorithm that you can access with a simple call. But if you want to compare and sort objects of the classes you make, you'll need to tell Python how to compare two of these objects by implementing the comparison dunder methods. Python calls the comparison dunder methods behind the scenes whenever your objects are used in an expression with the <, >, <=, >=, ==, and != comparison operators.

Before we explore the comparison dunder methods, let's examine six functions in the operator module that perform the same operations as the six comparison operators. Our comparison dunder methods will be calling these functions. Enter the following into the interactive shell.

```
>>> import operator
>>> operator.eq(42, 42)          # "EQual", same as 42 == 42
True
>>> operator.ne('cat', 'dog')   # "Not Equal", same as 'cat' != 'dog'
True
>>> operator.gt(10, 20)          # "Greater Than ", same as 10 > 20
False
>>> operator.ge(10, 10)          # "Greater than or Equal", same as 10 >= 10
True
>>> operator.lt(10, 20)          # "Less Than", same as 10 < 20
True
>>> operator.le(10, 20)          # "Less than or Equal", same as 10 <= 20
True
```

The operator module gives us function versions of the comparison operators. Their implementations are simple. For example, we could write our own operator.eq() function in two lines:

```
def eq(a, b):
    return a == b
```

It's useful to have a function form of the comparison operators because, unlike operators, functions can be passed as arguments to function calls. We'll be doing this to implement a helper method for our comparison dunder methods.

First, add the following to the start of *wizcoin.py*. These imports give us access to the functions in the operator module and allow us to check whether the other argument in our method is a sequence by comparing it to collections.abc.Sequence:

```
import collections.abc
import operator
```

Then add the following to the end of the *wizcoin.py* file:

```
--snip--
❶   def _comparisonOperatorHelper(self, operatorFunc, other):
        """A helper method for our comparison dunder methods."""

❷       if isinstance(other, WizCoin):
            return operatorFunc(self.total, other.total)
❸       elif isinstance(other, (int, float)):
            return operatorFunc(self.total, other)
❹       elif isinstance(other, collections.abc.Sequence):
            otherValue = (other[0] * 17 * 29) + (other[1] * 29) + other[2]
            return operatorFunc(self.total, otherValue)
        elif operatorFunc == operator.eq:
            return False
        elif operatorFunc == operator.ne:
            return True
        else:
            return NotImplemented
```

```
     def __eq__(self, other):  # eq is "EQual"
❺        return self._comparisonOperatorHelper(operator.eq, other)

     def __ne__(self, other):  # ne is "Not Equal"
❻        return self._comparisonOperatorHelper(operator.ne, other)

     def __lt__(self, other):  # lt is "Less Than"
❼        return self._comparisonOperatorHelper(operator.lt, other)

     def __le__(self, other):  # le is "Less than or Equal"
❽        return self._comparisonOperatorHelper(operator.le, other)

     def __gt__(self, other):  # gt is "Greater Than"
❾      return self._comparisonOperatorHelper(operator.gt, other)

     def __ge__(self, other):  # ge is "Greater than or Equal"
❿      return self._comparisonOperatorHelper(operator.ge, other)
```

Our comparison dunder methods call the _comparisonOperatorHelper()
method ❶ and pass the appropriate function from the operator module for
the operatorFunc parameter. When we call operatorFunc(), we're calling the
function that was passed for the operatorFunc parameter—eq() ❺, ne() ❻,
lt() ❼, le() ❽, gt() ❾, or ge() ❿—from the operator module. Otherwise, we'd
have to duplicate the code in _comparisonOperatorHelper() in each of our six
comparison dunder methods.

NOTE *Functions (or methods) like _comparisonOperatorHelper() that accept other functions as
arguments are called* higher-order functions.

Our WizCoin objects can now be compared with other WizCoin objects ❷,
integers and floats ❸, and sequence values of three number values that
represent the galleons, sickles, and knuts ❹. Enter the following into the
interactive shell to see this in action:

```
>>> import wizcoin
>>> purse = wizcoin.WizCoin(2, 5, 10)  # Create a WizCoin object.
>>> tipJar = wizcoin.WizCoin(0, 0, 37) # Create another WizCoin object.
>>> purse.total, tipJar.total # Examine the values in knuts.
(1141, 37)
>>> purse > tipJar # Compare WizCoin objects with a comparison operator.
True
>>> purse < tipJar
False
>>> purse > 1000 # Compare with an int.
True
>>> purse <= 1000
False
>>> purse == 1141
True
>>> purse == 1141.0 # Compare with a float.
True
>>> purse == '1141' # The WizCoin is not equal to any string value.
False
```

```
>>> bagOfKnuts = wizcoin.WizCoin(0, 0, 1141)
>>> purse == bagOfKnuts
True
>>> purse == (2, 5, 10) # We can compare with a 3-integer tuple.
True
>>> purse >= [2, 5, 10] # We can compare with a 3-integer list.
True
>>> purse >= ['cat', 'dog'] # This should cause an error.
Traceback (most recent call last):
  File "<stdin>", line 1, in <module>
  File "C:\Users\Al\Desktop\wizcoin.py", line 265, in __ge__
    return self._comparisonOperatorHelper(operator.ge, other)
  File "C:\Users\Al\Desktop\wizcoin.py", line 237, in _
comparisonOperatorHelper
    otherValue = (other[0] * 17 * 29) + (other[1] * 29) + other[2]
IndexError: list index out of range
```

Our helper method calls isinstance(other, collections.abc.Sequence) to see whether other is a sequence data type, such as a tuple or list. By making WizCoin objects comparable with sequences, we can write code such as purse >= [2, 5, 10] for a quick comparison.

SEQUENCE COMPARISONS

When comparing two objects of the built-in sequence types, such as strings, lists, or tuples, Python puts more significance on the earlier items in the sequence. That is, it won't compare the later items unless the earlier items have equal values. For example, enter the following into the interactive shell:

```
>>> 'Azriel' < 'Zelda'
True
>>> (1, 2, 3) > (0, 8888, 9999)
True
```

The string 'Azriel' comes before (in other words, is less than) 'Zelda' because 'A' comes before 'Z'. The tuple (1, 2, 3) comes after (in other words, is greater than) (0, 8888, 9999) because 1 is greater than 0. On the other hand, enter the following into the interactive shell:

```
>>> 'Azriel' < 'Aaron'
False
>>> (1, 0, 0) > (1, 0, 9999)
False
```

The string 'Azriel' doesn't come before 'Aaron' because even though the 'A' in 'Azriel' is equal to the 'A' in 'Aaron', the subsequent 'z' in 'Azriel' doesn't come before the 'a' in 'Aaron'. The same applies to the tuples (1, 0, 0)

(continued)

---

and (1, 0, 9999): the first two items in each tuple are equal, so it's the third items (0 and 9999, respectively) that determine that (1, 0, 0) comes before (1, 0, 9999).

This forces us to make a design decision about our WizCoin class. Should WizCoin(0, 0, 9999) come before or after WizCoin(1, 0, 0)? If the number of galleons is more significant than the number of sickles or knuts, WizCoin(0, 0, 9999) should come before WizCoin(1, 0, 0). Or if we compare objects based on their values in knuts, WizCoin(0, 0, 9999) (worth 9,999 knuts) comes after WizCoin(1, 0, 0) (worth 493 knuts). In *wizcoin.py*, I decided to use the object's value in knuts because it makes the behavior consistent with how WizCoin objects compare with integers and floats. These are the kinds of decisions you'll have to make when designing your own classes.

There are no *reflected* comparison dunder methods, such as __req__() or __rne__(), that you'll need to implement. Instead, __lt__() and __gt__() reflect each other, __le__() and __ge__() reflect each other, and __eq__() and __ne__() reflect themselves. The reason is that the following relationships hold true no matter what the values on the left or right side of the operator are:

- purse > [2, 5, 10] is the same as [2, 5, 10] < purse
- purse >= [2, 5, 10] is the same as [2, 5, 10] <= purse
- purse == [2, 5, 10] is the same as [2, 5, 10] == purse
- purse != [2, 5, 10] is the same as [2, 5, 10] != purse

Once you've implemented the comparison dunder methods, Python's sort() function will automatically use them to sort your objects. Enter the following into the interactive shell:

```
>>> import wizcoin
>>> oneGalleon = wizcoin.WizCoin(1, 0, 0) # Worth 493 knuts.
>>> oneSickle = wizcoin.WizCoin(0, 1, 0)  # Worth 29 knuts.
>>> oneKnut = wizcoin.WizCoin(0, 0, 1)    # Worth 1 knut.
>>> coins = [oneSickle, oneKnut, oneGalleon, 100]
>>> coins.sort() # Sort them from lowest value to highest.
>>> coins
[WizCoin(0, 0, 1), WizCoin(0, 1, 0), 100, WizCoin(1, 0, 0)]
```

Table 17-3 contains a full listing of the available comparison dunder methods and operator functions.

Table 17-3: Comparison Dunder Methods and operator Module Functions

Dunder method	Operation	Comparison operator	Function in operator module
__eq__()	**EQ**ual	==	operator.eq()
__ne__()	**N**ot **E**qual	!=	operator.ne()

Dunder method	Operation	Comparison operator	Function in operator module
__lt__()	**L**ess **T**han	<	operator.lt()
__le__()	**L**ess than or **E**qual	<=	operator.le()
__gt__()	**G**reater **T**han	>	operator.gt()
__ge__()	**G**reater than or **E**qual	>=	operator.ge()

You can see the implementation for these methods at *https://autbor.com/ wizcoinfull*. The full documentation for the comparison dunder methods is in the Python documentation at *https://docs.python.org/3/reference/datamodel .html#object.__lt__*.

The comparison dunder methods let objects of your classes use Python's comparison operators rather than forcing you to create your own methods. If you're creating methods named equals() or isGreaterThan(), they're not Pythonic, and they're a sign that you should use comparison dunder methods.

Summary

Python implements object-oriented features differently than other OOP languages, such as Java or C++. Instead of explicit getter and setter methods, Python has properties that allow you to validate attributes or make attributes read-only.

Python also lets you overload its operators via its dunder methods, which begin and end with double underscore characters. We overload common mathematical operators using the numeric and reflected numeric dunder methods. These methods provide a way for Python's built-in operators to work with objects of the classes you create. If they're unable to handle the data type of the object on the other side of the operator, they'll return the built-in NotImplemented value. These dunder methods create and return new objects, whereas the in-place dunder methods (which overload the augmented assignment operators) modify the object in-place. The comparison dunder methods not only implement the six Python comparison operators for objects, but also allow Python's sort() function to sort objects of your classes. You might want to use the eq(), ne(), lt(), le(), gt(), and ge() functions in the operator module to help you implement these dunder methods.

Properties and dunder methods allow you to write classes that are consistent and readable. They let you avoid much of the boilerplate code that other languages, such as Java, require you to write. To learn more about writing Pythonic code, two PyCon talks by Raymond Hettinger expand on these ideas: "Transforming Code into Beautiful, Idiomatic Python" at *https://youtu.be/OSGv2VnC0go/* and "Beyond PEP 8—Best Practices for Beautiful, Intelligible Code" at *https://youtu.be/wf-BqAjZb8M/* cover some of the concepts in this chapter and beyond.

There's much more to learn about how to use Python effectively. The books *Fluent Python* (O'Reilly Media, 2021) by Luciano Ramalho and *Effective Python* (Addison-Wesley Professional, 2019) by Brett Slatkin provide more in-depth information about Python's syntax and best practices, and are must-reads for anyone who wants to continue to learn more about Python.

INDEX

Numbers and Symbols

256 objects and 257 objects, 154–155

./, using with Ubuntu, 42

/? command line argument, 25–26

= assignment operator, 113

== comparison operator, 113, 336
chaining, 103, 105
using to compare objects, 154
using with None, 94–95

!= comparison operator, 336

* and ** syntax
using to create variadic functions, 167–171
using to create wrapper functions, 171–172
using with arguments and functions, 166–167

* character, using as wildcard, 28–29

? character, using as wildcard, 28–29

[:] syntax, using, 104

< comparison operator, 337

<= comparison operator, 337

> comparison operator, 337

>= comparison operator, 337

-> (arrow), using with type hints, 191

\ (backslash)
purpose of, 18
using with strings, 95

: (colon), using with lists, 97–98, 104

, (comma), including in single-item tuples, 150

- (dash), using with command line arguments, 25

$ (dollar sign), using in macOS, 23

. (dot), using with commands, 31

-- (double dash), using with command line arguments, 25

__ (double underscore), using in dunder methods, 322. *See also* underscore (_)

/ (forward slash)
purpose of, 18
using with command line arguments, 25

(hash mark)
using with comments, 183
using with docstrings, 188

[] (index operator), using, 117

; (semicolons), using with timeit module, 226–227

' (single quote), using, 46

~ (tilde), using in macOS, 23

- (unary operator), 155–157

+ (unary operator), 156–157

_ (underscore)
PEP 8's naming conventions, 60–61
as prefix for methods and attributes, 291–292
private prefix, 81
using with _spaces attribute, 290
using with dunder methods, 120
using with private attributes and methods, 283
using with WizCoin class, 279

A

abcTraceback.py program, saving, 4

__abs__() numeric dunder method, 328

absolute versus relative paths, 20–21

__add__() numeric dunder method, 327

addToTotal() function, creating, 172–173

C

C:\ part of path, 18

-c switch, using to run code from command line, 26

callables and first-class objects, 121–122

camelCase, 60

casing styles, 60

Catalina version, 23

cd (change directories) command, 29–30

__ceil__() numeric dunder method, 328

chaining operators, 103, 105, 159–160

child class, creating, 294

class attributes, 306

class methods, 304–306

class objects, 284

classes. *See also* inheritance

    creating, 77

    creating objects from, 278

    creating WizCoin, 279–284

    defined, 276

    designing for real world, 290–291

    as functions or modules, 77

    "is a" relationships, 299

clause vs. block vs. body, 123–124

CLI (command line interface), 22

close() and open() functions, 93–94

cls and clear (clear terminal) commands, 35

code. *See also* source code

    avoiding guesses, 90

    beauty of, 88

    commented out, 74–75

    flat vs. nested, 89

    formatting for readability, 12–13

    implementation, 90

    interrupting, 134

    namespaces, 91

    organizing, 77

    readability of, 89

    running from command line, 26

    silenced errors, 89–90

    simplicity and complexity of, 89

    sparse vs. dense, 89

    special cases, 89

    speed of, 90

    verbose and explicit, 89

code formatting, defined, 45

code point, getting for characters, 146–147

code smells. *See also* source code

    classes as functions or modules, 77

    commented-out code, 74–75

    dead code, 74–75

    defined, 69

    duplicate code, 70–71

    empty except blocks, 79–80

    error messages, 79–80

    list comprehensions, 77–79

    magic numbers, 71–73

    myths, 80–84

    print debugging, 75–76

    summary, 84–85

    variables with numeric suffixes, 76

codetags and TODO comments, 187

coercion, explained, 128

collections module, contents of, 120

collections.defaultdict, using for default values, 99–100

colon (:), using with lists, 97–98, 104

comma (,), including in single-item tuples, 150

command history, viewing, 28

command line

    arguments, 24–26

    options, 25

    overview, 22–23, 42

    running code with -c switch, 26

    running programs from, 23–24

    running *py.exe* program, 26–27

    running Python programs from, 26

    tab completion, 27–28

    terminal window, 23

Command Prompt shell, 23

commands

    canceling, 28

    cd (change directories), 29–30

    cls and clear (clear terminal), 35

    copy and cp (copy files), 31–32

    del (delete files and folders), 33–34

    dir (list folder contents), 30

    dir /s (list subfolder contents), 31

    find (list subfolder contents), 31

    ls (list folder contents), 30

git add command, 223
git clone command, 223
git commit command, 223
git diff, using, 211–213
git filter-branch command, 220
git init command, 223
GitHub and git push command,
        221–223
glob patterns, explained, 29
global variables, myth about, 82–83
glossary, accessing, 108, 131
GrandchildClass, creating, 294–295
Greater Than operation, 337
Greater than or Equal operation, 337
__gt__() comparison dunder
        method, 337
GUI (graphical user interface), 22
GUI Git tools, installing, 203–204

H

hash mark (#)
    using with comments, 183
    using with docstrings, 188
hashes, defined, 117–119
--help command line argument, 25–26
help with programming, asking for,
        9–14
higher-order functions, 174. *See also*
        functions
home directory, 19
Homebrew, installing and
        configuring, 213
horizontal spacing, 47–51
Hungarian notation, 63

I

id() function, calling, 111, 154
identifiers, defined, 59
identities, defined, 111–114
IEEE 754 standard and floating-point
        numbers, 147–148
if statement as clause header, 124
immutable and mutable objects,
        114–117, 144
in operator, using with values of
        variables, 105
increment and decrement operators,
        156–157

indentation, using space characters
        for, 47–48. *See also* significant
        indentation
index operator ([ ]), using, 117–118
index() string method, exception
        related to, 179
indexes, defined, 117–119
inequality != operators, avoiding
        chaining, 149–150
inheritance. *See also* classes;
        multiple inheritance;
        OOP (object-oriented
        programming)
    base classes, 296
    best practice, 308–309
    class attributes, 306–307
    class methods, 304–306
    vs. composition, 299–301
    creating child classes, 294
    derived classes, 296
    downside, 301–303
    explained, 293
    isinstance() and issubclass()
        functions, 303–304
    MRO (method resolution order),
        310–312
    overriding methods, 296–297
    static methods, 306–307
    subclasses, 296
    super classes, 296
    super() function, 297–299
__init__(), and self, 280–282
__init__.py file and packages, 121
inline comments, 183–184
in-place augmented assignment dunder
        methods, 330–332
installing
    Black code formatting tool, 54
    Git, 202–204
    Homebrew, 213
    Meld for Linux, 213
    Mypy, 192–193
    Pyflakes, 9
    tkdiff, 213
instances, defined, 111–114, 276. *See
        also* isinstance()
instruction set, explained, 129
int() function, using, 158

properties
- vs. attributes, 128–129
- best practices, 322
- read-only, 320–321
- turning attributes into, 316–319
- using, 316

public access attributes and methods, 283

pure function, 173–174. *See also* functions

push command, using in Git, 221–223

.py source code files, locating, 200

.pyc files, bytecode in, 129

py.exe program, running, 26–27

Pyflakes, installing, 9

PyPy just-in-time compiler, 108

Python
- documentation, 121
- error messages, 4–8
- glossary, 108, 131
- language and interpreter, 108–109
- programming language, 109

Python programs, running without command line, 39–42. *See also* programs; *The Zen of Python*

Python Standard Library, 120–121. See also library vs. framework vs. SDK vs. engine vs. API

pythonic code, core of, 104

Q

__qualname__ attribute and type() function, 284–285

questions, asking, 10–11, 14–15

R

__radd__() reflected numeric dunder method, 330

raising exceptions, 90, 178–179

__rand__() reflected numeric dunder method, 330

range() vs. enumerate(), 92–93, 103–104

rd and rmdir commands, 34–35

__rdivmod__() reflected numeric dunder method, 330

readlines() and open() functions, using, 126

README files, 200, 211–212, 215–216, 218

read-only properties, 320–321

RecursionError exception, raising, 318–319

references, explained, 137–138

reflected numeric dunder methods, 328–330

relative vs. absolute paths, 20–21

renaming files and folders, 32–33

repo. *See also* Git
- cloning for GitHub repo, 222–223
- creating, 223
- creating on computer, 206–207
- deleting and moving files in, 215–216
- deleting files from, 214–215
- ignoring files in, 209–210
- and version control systems, 200

__repr__() method, using, 325

repr string, sensitive information in, 325

requests module, *sessions.py* file, 188–189

return values and data types, 177–178. *See also* values

__rfloordiv__() reflected numeric dunder method, 330

__rlshift__() reflected numeric dunder method, 330

rm (removing files and folders) command, 33–34

__rmatmul__() reflected numeric dunder method, 330

__rmod__() reflected numeric dunder method, 330

__rmul__() reflected numeric dunder method, 330

roll back, performing in Git, 217–220

root folder, explained, 18

__ror__() reflected numeric dunder method, 330

__round__() numeric dunder method, 328

__rpow__() reflected numeric dunder method, 330

__rrshift__() reflected numeric dunder method, 330

__rshift__() numeric dunder method, 328

Beyond the Basic Stuff with Python is set in New Baskerville, Futura, Dogma, and TheSansMono Condensed. The book was printed and bound by Sheridan Books, Inc. in Chelsea, Michigan. The paper is 60# Finch Offset, which is certified by the Forest Stewardship Council (FSC).

The book uses a layflat binding, in which the pages are bound together with a cold-set, flexible glue and the first and last pages of the resulting book block are attached to the cover. The cover is not actually glued to the book's spine, and when open, the book lies flat and the spine doesn't crack.

RESOURCES

Visit *nostarch.com/beyond-basic-stuff-python/* for errata and more information.

More no-nonsense books from **NO STARCH PRESS**

AUTOMATE THE BORING STUFF WITH PYTHON, 2ND EDITION
Practical Programming for Total Beginners
BY AL SWEIGART
592 PP., $39.95
ISBN 978-1-59327-992-9

PYTHON CRASH COURSE, 2ND EDITION
A Hands-On, Project-Based Introduction to Programming
BY ERIC MATTHES
544 PP., $39.95
ISBN 978-1-59327-928-8

PYTHON ONE-LINERS
Write Concise, Eloquent Python Like a Professional
BY CHRISTIAN MAYER
216 PP., $39.95
ISBN 978-1-7185-0050-1

SERIOUS PYTHON
Black-Belt Advice on Deployment, Scalability, Testing, and More
BY JULIEN DANJOU
240 PP., $39.95
ISBN 978-1-59327-878-6

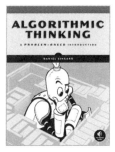

ALGORITHMIC THINKING
A Problem-Based Introduction
BY DANIEL ZINGARO
408 PP., $49.95
ISBN 978-1-7185-0080-8

REAL WORLD PYTHON
A Hacker's Guide to Solving Problems with Code
BY LEE VAUGHAN
360 PP., $34.95
ISBN 978-1-7185-0062-4

PHONE:
800.420.7240 OR
415.863.9900

EMAIL:
SALES@NOSTARCH.COM

WEB:
WWW.NOSTARCH.COM

The Electronic Frontier Foundation (EFF) is the leading organization defending civil liberties in the digital world. We defend free speech on the Internet, fight illegal surveillance, promote the rights of innovators to develop new digital technologies, and work to ensure that the rights and freedoms we enjoy are enhanced — rather than eroded — as our use of technology grows.

EFF.ORG

ELECTRONIC FRONTIER FOUNDATION

Protecting Rights and Promoting Freedom on the Electronic Frontier